What Kind of Regionalism?

Josefina Syssner

What Kind of Regionalism?

Regionalism and Region Building in Northern European Peripheries

PETER LANG
Frankfurt am Main · Berlin · Bern · Bruxelles · New York · Oxford · Wien

Bibliographic Information published by Die Deutsche Bibliothek
Die Deutsche Bibliothek lists this publication in the Deutsche Nationalbibliografie; detailed bibliographic data is available in the internet at <http://dnb.ddb.de>.

ISBN 3-631-55201-7
US-ISBN 0-8204-9898-X

www.peterlang.de

Contents

Acknowledgements

For as long as I can remember, I have harboured a desire to write. At the age of six or seven, I started composing what with a little goodwill could be called detective novels. The novels were always composed according to a thoroughly set pattern. There were only a few words per page, and the illustrations were always many and detailed. In nine cases out of ten, the story was about some young girl who had her pony stolen, got it back again and then celebrated the triumph with great joy and jubilation. The audience for these novels consisted solely of my younger sister Johanna, who was roused from her fundamental lack of interest in my efforts only when she asked if I couldn't think about writing about a dog for once, 'just for a change'.

As a 10 or 12 year old, I had left the detective novels behind and was determined to become a poet. One of my deepest beliefs during this period was that a poet has a choice of two equally emotional genres: the kind of poetry that is melodramatic and sentimental; and the romantic, passionate variety that sings the praises of the beauty of nature. No matter what the genre, I thought, all poetry is about piling emotional rhymes on top of each other. My first test as a poet came in the spring of 1986. The schoolmaster had decided that the whole school should gather for something that seemed to be a cross between a commemoration for Olof Palme, the murdered Prime Minister, and an attempt to welcome the eagerly awaited spring. One could indeed object that these two aims were slightly incompatible. That, however, bothered neither me nor my friend Linda when we were each asked to write a poem to be read out at this very ceremony. For myself, I was actually quite delighted. I had already practised writing melodramatic elegies and composing joyful poems about purling rills and the green of spring. Now I was ready for my breakthrough as the village poet. But when the time came for our 15 minutes of fame, Linda and I completely lost our professional nerve. Instead of reading our well-prepared, moving poems, we burst out laughing at both Palme and the much longed-for spring, until the schoolmaster sternly ordered us off the stage. Since then, my interest in poetry has been very modest, mostly because the art of poetry requires one to take oneself, as well as the result of one's efforts, very seriously.

After my wrecked attempts to write detective stories and poetry, I acquired an interest in journalism. After an educational visit to the local newspaper, I decided to become a managing editor. I visualized myself reorganizing the work at the old, mildewy editorial office. I saw myself producing grandiose news about this and that, and I dreamt about seeing my name in the paper every day. When at the age of 14 I spent a week as a trainee at the local newspaper, I realized that it was strictly prohibited to write articles that did not deal with sports, did not flatter the local government commissioner, or was longer than a short paragraph. After that, my interest in journalism waned considerably.

During all those years, not once do I recall having uttered the word 'university'. As far as I knew, no one in my milieu had ever studied at university. The academic world played no role at all in our everyday life, and accordingly we had little use for such a word. My schoolteachers, in all likelihood, would have had some sort of experience of higher education; yet I can't recall any of them ever talking about the possibility that we might continue our studies after leaving school. Instead, as it happened, it was my friend Nea who, right after we turned 18, asked me if I was planning to go to university. My first thought was that it was expensive, that one had to take out loans, and that I was anyhow far too stupid. However, Nea showed me a copy of the handbook of Lund University, listing all the courses. Most course titles didn't mean anything to me, but one of them – 'Social Sciences Programme' – sounded straightforward enough for me to dare to apply. Six months later I received a letter of acceptance from the university. I packed my father's truck with furniture, drove the 350 kilometres to Lund and attended the introductory meeting, still with a truck-load of furniture outside.

The first semesters at the university were mostly about learning things by heart. Together with my friends Anna and Margareta and a hundred other students in our course, I learned to rattle off the constitution of Tanzania, basic Swedish commercial law and other essentials. We learned how to calculate the likelihood of the most obscure events, and to summarize the difference between positivism and hermeneutics. I wouldn't say that it was boring, but it was, to say the least, not very creative. After a year or so, I ended up at the Department of Social and Economic Geography. After taking some courses there, for the first time since I came to Lund I had a vague sense that the essence of a university education wasn't to learn things by heart. In a course on 'Human Geography and the Theory of Science', our teacher let us discuss scientific concepts and basic methodological issues. The idea of this course was not that we should cram our heads full of 'knowledge' and regurgitate it in the written exam. Instead, we were encouraged to reflect on and even critically discuss what we had learned. Evidently, I never dared to tell the teacher (whose name was P.O. Hallin, by the way), but it was at this point that for the first time I found university education wholly exciting, absorbing and fun.

Around the same time, I grasped that it is writing things, not memorizing them, that is central to the social sciences. Besides, I realized that I could write as I really wanted to. The main point of writing was no longer to be funny or exciting. Nobody forced me to write about dogs, local commissioners or sports. I didn't have to allude to the emotional life of the reader, and I didn't have to do rhymes or express myself in short paragraphs.

It was with this cheerful attitude to life and to university that, some years later, I enrolled as a postgraduate student at Linköping University. The opportunity to spend several years reading, writing and reflecting was, as I saw it, a huge privilege. The feeling of being privileged has followed me ever since. And I still enjoy writing. The problem (and it is at this point that I start spreading around sincere apologies and warm thanks) is just that it is not always fun to read what I write. Hence, after a few years as a Ph.D. candidate, I decided to start writing in English. This decision was taken against all reason, and meant that both my supervisors have been subjected to superhuman strain over the last few years.

Professor Rune Johansson, my main supervisor, deserves my deepest and most sincere thanks. During my years as his Ph.D. student, Rune has gradually raised the standards and demanded more clarity and consistency. Throughout the process, he has worked as a sounding board and as an encouraging discussant. Quite apart from his intellectual capacity, Rune has showed consideration and kindness that meant a lot to me when the work on my thesis was heavy going. Besides, Rune has been patient enough to correct my mistaken use of the words 'have' and 'has' – over and over again. Thanks for that, Rune!

Associate Professor Peo Hansen has been my assistant supervisor. Peo has the ability to come up with sharp and critical comments at exactly the right moment and, at equally right moments, to come up with support, encouragement and concrete advice. His supervision has always been stimulating and fruitful. Thanks Peo, not least for your attempts to explain, in a very humble and diplomatic way indeed, that I might think about expanding my English vocabulary somewhat, and dispense with words I invented for myself that were hitherto unknown in the English language.

Besides Rune and Peo, I want to thank my colleagues at the Department for Ethnic Studies. I have a particular feeling of gratitude towards Magnus Dahlstedt, Christina Johansson, Anna Lundberg, Riikka Norrbacka-Landsberg, Erik Olsson, Tünde Puskás, Jens Stillhoff-Sörensen and Henning Süssner, all of whom read and commented on my texts at different stages of the writing process. I would also like to thank Slave Saveski and Hamid Gharakhani for helping me with my computer problems, always at lightning speed and with a kindly smile.

Thanks also to Richard Ek, Eric Clark, Christine Hudson and Katarina Schough for their generous comments on the first drafts of the manuscript.

One of my years as a postgraduate was spent at the Department of Political Sciences at Rostock University, Germany. One of those who contributed to making my stay in Rostock both worthwhile and pleasant is Professor Nikolaus Werz, who really made a lot of effort to involve me in the activities at the department. Thank you, Professor Werz!

In recent years I have been fortunate enough to meet and to talk to people involved in the ECPR's Standing Group on Regionalism. One of those who contributed with his experience, in a very generous and open-minded way, is Jörg Mathias, the Academic Secretary of the Standing Group. Thank you, Jörg, for your encouragement! Another person who ranks high among my top-ten favourite people right now is Michael James, who copy-edited the entire manuscript. Thank you so much for your help, and sorry about all the headaches my texts must have caused you! I would also like to thank Jesper Weithz, who helped me with the graphic design of the final manuscript, and Ute Winkelkoetter, who has been my very kind and skilful contact at Peter Lang Publishing Group. Thanks also to *Lars Hiertas Minnesfond*, *Helge Ax:son Johnsons Stiftelse* and German Academic Exchange Service (DAAD) for their financial support during this project.

In these very last days at work, however, my deepest and warmest thoughts go to my family, who filled my private life with so much meaning during these years. I want to thank my parents, my brothers and sister for the sense of confidence and joy you assisted with over the years. Thanks also to Henning for support and encouragement. Lastly, I want to express my never-ending gratitude for having the privilege of living my life together with our lovely kids. Ivan, Ella and yet unborn Baby X – you really are the sunshine of my life! Nothing in this world makes me happier than you do. I hope that, some fine day, I will get the opportunity to write a book that is funnier for you kids to read.

Josefina Syssner,
Norrköping, December 2005

Introduction

What kind of regionalism is emerging in less favoured, politically defined regions in Europe of today? Contemporary research provides little guidance on the matter. The study of regions, regionalism and regionalization has indeed been regarded a 'boom industry', and the idea that we are moving towards a 'Europe of the regions' has been described as 'both an intellectual and political fashion'.[1] Much of the literature on regionalism, however, revolves around either 'ethnic' regions or successful growth regions. In many recent volumes on European regions, regionalism or regionalization, we learn primarily about 'successful' regions, 'bourgeois' regions, and regions that play an ever-increasing role in the European policy process. When this is not the case, the focus is rather on political mobilization in regions known for hosting 'substate nationality claims'[2] on the basis of ethnic identities. Regionalism in regions that are neither economically successful nor regarded as 'ethnic communities' does not seem to catch the interest of the regional sciences today.

Many accounts of regionalism in western Europe revolve around narratives about old and new regionalism. *Old* regionalism is largely understood as a movement of the 1960s and 70s, described 'principally in terms of ethnonationalistic movements',[3] and located primarily on the periphery. Ethnoregional mobilization has partly been interpreted as a result of uneven economic development combined with cultural differences between the centre and the periphery.[4] The examples of the Basque Country, Corsica and Brittany have been

1 Le Galès, P. and Lequesne, C. (1998), 'Series Editors Preface' in *Regions in Europe*, New York: Routledge, p.vii; see also Keating, Michael and Loughlin, John (eds.) (1997), *The Political Economy of Regionalism*, London: Frank Cass, p.1

2 Keating, Michael (2004), 'European Integration and the Nationalities Question', *Politics & Society*, 32(3), pp. 367-388, p.368.

3 Schmitt-Egner, Peter (2002), 'The Concept of 'Region': Theoretical and Methodological Notes on its Reconstruction', *European Integration*, 24(3), pp.179-200, p.188. See also Gidlund, Janerik and Sörlin, Sverker (1993), *Det Europeiska Kalejdoskopet*, Stockholm: SNS Förlag, p.161.

4 This was maintained by Hechter, Michael (1975), *Internal Colonialism*, Berkeley, CA: University of California Press, p.43. For a critical discussion on that, see Olzak, Susan (1998), 'Ethnic Protest in Core and Periphery States', *Ethnic and Racial Studies*, 21(2), pp.187-217 and Keating, Michael (1998), *The New Regionalism in*

cited to illustrate this form of old regionalism. Old regionalism, furthermore, is said to have been distinguished by its social and cultural agenda and by its hostile attitude towards the central state.[5]

Old regionalism is currently said to have been largely (but not completely) replaced with a *new* European regionalism which in many respects is narrated as a manifest opposite to the old one.[6] New regionalism is said to be driven not by social or cultural demands, but by economic self-interest.[7] And whereas old regionalism is said to have involved social mass movements, new regionalism has been associated primarily with regional functionaries involved in regional development coalitions, partnerships and networks.[8] Studies on new regionalism have, furthermore, been focused, not on ethnic communities on the periphery, but on growth regions in the economic and political centres of western Europe. These affluent regions are thought to benefit from a greater degree of economic autonomy, since, at least at first sight, they lose rather than gain from the spatial reallocation of resources.[9]

Incidentally, it must be observed that the concepts of both old and new regionalism also have been used to describe political and scholarly approaches to economic development.[10] The 'old' way of handling regional development

---

Western Europe – Territorial Restructuring and Political Change, Cheltenham UK, Northampton, MA USA: Edward Elgar, p.19.

5 See Gren, Jörgen (2002a), 'New Regionalism and West Sweden: The Factors of Change in the Regionalism Paradigm', *Regional and Federal Studies* , 12(3), pp.79 – 101, p.81.

6 Törnqvist, Gunnar (1993), *Sverige i Nätverkens Europa*. Malmö: Liber-Hermods, p.42; Gren, Jörgen (1999), *The New Regionalism in the EU: the lessons to be drawn from Catalonia, Rhône-Alpes and West Sweden*, Stockholm: Fritzes; Gren, Jörgen (2002b), *Den Perfekta Regionen*, Lund: Studentlitteratur, pp.16ff.

7 Rumford, Chris (2002), *The European Union - A Political Sociology*, Oxford: Blackwell, p.178; Cole, Alistair and Loughlin, John (2003), 'Beyond the Unitary state? Public Opinion, Political Institutions and Public Policy in Brittany', *Regional Studies*, 37(3), pp.265-276., p.266.

8 See Gren (2002b:33); Sörlin and Gidlund (1993:219).

9 For a discussion on this, see Keating, M., Loughlin, J. and Deschouwer, K. (2003) *Culture, Institutions and Economic Development*, Cheltenham, Northampton: Edward Elgar, p.7.

10 Lovering, John (1999), 'Theory Led by Policy: The Inadequacy of 'The New Regionalism', *International Journal of Urban and Regional Research*, 23(2), pp.379-395; MacLeod, Gordon (2001), 'New Regionalism Reconsidered: Globalization and the Remaking of Political Economic Space', *International Journal of Urban and Regional Research*, 25(4) pp.804-829; MacLeod, Gordon and Jones, Martin (2001), 'Renewing the Geography of Regions', *Environment and Planning*

issues, it is argued, was guided by a 'top-down regionalization approach'.[11] Regions were at the time seen primarily as '*objects* of policies designed by national governments to bring about a convergence of economically weaker regions with the most advanced'.[12] The 'new' approach to economic development has left much of the interventionist Keynesian ideal of the 1960s and 70s behind, and focuses instead on endogenous growth in all regions.[13] Regions are no longer seen as complementary parts in a national whole, but are thought to provide for their own growth and competitiveness, primarily through a focus on '"knowledge-intensive" innovation and flexibility'.[14] Like the more political versions of new regionalism, this perspective on economic development is understood as closely connected to globalization, Europeanization, and an alleged withdrawal, 'hollowing-out' or restructuring of the nation state.

What Kind of Regionalism?

The pertinence of describing old regionalism only in terms of ethnic movements located on the periphery may be disputed.[15] And we should not be too quick to draw conclusions about the actual differences between old and new regionalism. Yet many accounts of regionalism in western Europe do revolve around two types of regions: (*a*) ethnically defined regions with sub-nationalist demands (some of them located on the periphery), and (*b*) affluent regions in the economic and political centre of Europe.[16] Still, many regions in Europe qualify

---

D, 19(6), pp.669-695; Jones, Martin (2001), 'The Rise of the Regional State in Economic Governance: 'Partnerships for Prosperity' or New Scales of State Power?', *Environment and Planning A*, 33, pp.1185-1211; Ceccato, Vania and Persson, Lars Olof (1999), *Forskning för regional utveckling*, Stockholm: Nordregio.

11 Gren (2002a:80).

12 Cole and Loughlin (2003:265); Keating, Michael (2001), 'Rethinking the Region: Culture, Institutions and Economic Development in Catalonia and Galicia', *European Urban and regional Studies* 8(3), pp.217-234, p.217.

13 Lovering, John (2001), 'The Coming Regional Crisis (And How To Avoid It)', *Regional Studies*, 35(4), pp. 349-354, p.349.

14 Lovering (2001:349).

15 For a discussion on aspects of 'rurality' in ethno-nationalist conflicts, see Nairn, Tom (1997), *Faces of Nationalism*, London and New York: Verso, p.90ff. See also Olzak (1998:200).

16 For a similar discussion, see Taylor, Adam (2005), 'Bringing in the Norm: Non-ethnic, Non-Exemplar Regions', Paper presented at *Regional Identities, Cultures and Images – a Path to Regional Development?*, Norrköping.

neither as 'ethnic' communities nor as booming growth regions. One might indeed presume that the administratively or politically defined, less favoured regions of today host no regionalist tendencies whatsoever. There are, however, many indications to the contrary. Regions in Europe are said to be undergoing a renaissance in quite general terms, and regions are generally thought of as increasingly important for economic activities, political action and cultural identification.[17] Accordingly, we need to ask: *What kind of regionalism is emerging in less favoured, politically defined regions of today?* In this study, I wish to respond to that question by analysing policy documents from two political regions in the northern European periphery – Norrbotten in Sweden and Mecklenburg-Western Pomerania in Germany.

The Value-basis of Regionalism?

A central question in this work is how regionalism as such is to be studied. In previous research, regionalism is often referred to as a process, and much academic discussion has concerned the reasons for and the effects of regionalism. There is a wealth of theories on why regionalism arises, the circumstances under which it becomes popular, and what popular regionalism may bring in terms of political stability and economic development. These questions are not irrelevant or uninteresting, but they actually leave 'regionalism' – as a concept and as an object of study – more of a 'black box' than one could expect. For anyone who is interested in knowing more about the ideological foundations or value-basis of regionalism, remarkably little literature is to hand. For that reason, I would suggest that regionalism could be analysed as a set of ideas or even as an instance of a political ideology.

By equating regionalism with a political ideology, I do not mean that regionalism is 'false', 'wrong' or something that in all cases should be understood as opposite to 'reality'. Instead, I use the concept of ideology to describe a political

---

17 Törnqvist, Gunnar (1998), *Renässans för Regioner*, Stockholm: SNS Förlag; Gidlund and Sörlin (1993:161ff.); Harmaakorpi, Vesa (2004), *Building a Competitive Regional Innovative Environment – The regional Development Platform as a Tool For Regional Innovation Policy*, Helsinki: Helsinki University of Technology. See also: Hogwood, Patricia and Gomez, Ricardo (2004), 'Implementing European Policy in Post Devolution Britain: Lessons from other EU Member States', *Manchester Papers in Politics* (11), Manchester: Department of Government, University of Manchester; Clement, Keith and Hansen, Malin (2001), *Sustainable Regional Development in the Nordic Countries*, Stockholm: Nordregio.

belief-system, shared by a group of people. [18] Like most other political projects, regionalism is based on a set of norms, beliefs, demands and interests that are possible for us to observe.[19] This leads to the more specific question: *What is the value-basis of regionalism in less favoured, politically defined regions of today?* Thus, what the value-basis of regionalism would be in regions that are neither economically successful nor ethnically defined is far from given. The ethnic or cultural 'old regionalist' demands would presumably not be commonly found among representatives of administratively or politically defined regions. And claims for economic autonomy might win little sympathy in regions that over the years have been net recipients of state subsidies. So what is then understood as a 'regional interest' in such regions?

If we manage to identify the 'distinctive core'[20] of regionalism in one specific context, it would be worthwhile to carry out further research on regionalist ideologies in other places and times. An analysis of regionalist ideologies may also enrich our knowledge about ideologies in more general terms. Hence, in many textbooks on political ideology, *grand* ideologies such as conservatism, socialism or liberalism are the centre of attention. In cases where this triad of classical ideologies is complemented with accounts for modern cross-party ideologies, the examples mentioned are in most cases racism, nationalism, feminism or ecologism.[21] Taking into account that there is a lack of conclusive studies of regionalism-as-ideology (though regionalism is considered an important element in the Europe of today), this study contributes to partially filling a gap in our knowledge.

---

18 van Dijk, Teun A. (2004), 'Politics, Ideology and Discourse', in Wodak, R. (ed.), *Encyclopedia of Language and Linguistics*, Barcelona: Universitat Pompeu Fabra., p.2

19 For a discussion on regionalist' demands, see Keating, Michael (1997), 'The Political Economy of Regionalism' in Keating, M. and Loughlin, J. (eds.), *The Political Economy of Regionalism*, London: Frank Cass, p.18.

20 Bastow, S., Martin, J., and Pels, D. (2002), 'Introduction: Third Ways in Political Ideologies', *Journal of Political Ideologies*, 7(3), pp.269-280, p.269.

21 van Dijk (2004:10); Rejai, Mostafa (1995), *Political Ideologies, A Comparative Approach*, London and New York: M.E Sharpe; Funderburk, Charles and Thobaben, Robert G. (1989), *Political Ideologies. Left, Center, Right*, New York: Harper & Row Publishers; Ball, Terence and Dagger, Richard (1999), *Political Ideologies and the Democratic Ideal*, New York: Longman; Baradat, Leon P. (1991), *Political Ideologies. Their Origin and Impact*, New Jersey: Prentice Hall.

The Cases

In order to answer the questions posed above, I have studied two northern European regions – the Swedish county of Norrbotten and the German *Land* Mecklenburg-Western Pomerania (MV). Both Norrbotten and MV are politically defined regions, with their own political administrations, though with considerable differences in legal capacity. Both regions are distinguished by a relatively peripheral position in both economic and political terms. Mecklenburg-Western Pomerania is one of the so-called *Neuen Bundesländer*, and ever since the German unification economic problems, high unemployment rates, and net migration losses have impeded the development of the *Land*. Norrbotten, located in the northernmost part of Sweden, has higher unemployment than the Swedish average.[22] Compared with MV, the unemployment figures here are indeed surmountable. The structural transformations of Norrbottnic industry, the great distances within the county, and the low population density, however, position Norrbotten as a less favoured region.

Outline of the Book

The outline of this book is as follows. *Chapter 1* deals with the theoretical framework of the study. Concepts such as region, regionalism and region building are further discussed. The way in which I understand the concept of ideology is explained, and it will be suggested that insights from studies of nationalism would be of use in the study of regionalism. *Chapter 2* describes the most important methodological aspects of the study. This is also where I explain how the sources have been selected and analysed. Moreover, it is suggested that ideologies such as regionalism may very well be studied through an analysis of policy discourses. In *Chapter 3*, the case regions are introduced to the reader. Some central questions concerning the comparability of Norrbotten and Mecklenburg-Western Pomerania will also be posed. *Chapters 4–8* then present the empirical findings. Each chapter corresponds to one dimension of regionalist ideology in terms of which the cases of Norrbotten and Mecklenburg-Western Pomerania are compared and discussed. Finally, *Chapter 9* contains a concluding discussion and an attempt to answer the question of what kind of regionalism is evolving in less favoured, politically defined regions.

---

22 Glesbygdsverket (2002), 'Kommuner med den lägsta sysselsättningen. 2001. Andel sysselsatta. Procent' at *www.glesbygdsverket.se.*

1. Regions, Regionalisms and Region Building: A Theoretical Framework

In *The Political Economy of Regionalism*, Michael Keating points out that regionalism in western Europe covers a set of phenomena so complex and diverse that 'some observers have doubted whether the concept has any analytical value at all'.[23] One example of the ambiguity of the concept is that regionalism has been understood not only as a political strategy from below, but also as a top-down governmental strategy.[24] In its latter form, regionalism would be reflected in extensive direct subsidies and public investments steered to less favoured regions by the central government. These actions would then be part of a governmental policy aiming at promoting a spatial reallocation of resources, and at preventing marginalization of the periphery.[25] Regionalism, when understood in this way, is then to be described as a political strategy which 'derives from a "modernist" theory of nation-building and maintenance'.[26]

In this study, the focus will be on regionalism and regionalist policy *within* the regions of Norrbotten and Mecklenburg-Western Pomerania. Even when regionalism is understood as an intra-regional phenomenon, however, there is a conceptual confusion. In some cases, regionalism is referred to, investigated and discussed without being properly defined at the first stage. In other cases, we find several competing definitions. Regionalism has, for instance, been defined

---

23 Keating (1997:17); see also Castensson, Reinhold (1994), 'Region – ett mångfaldigt begrepp', in Blomberg, B. and Lindquist, S.O (eds.), *Den Regionala Särarten*, Lund: Studentlitteratur, p.48

24 The concept of regionalism has also been used to refer to the establishment of international trading blocs in the global economy. EU, NAFTA and ASEAN have then been cited as manifestations of this kind of (meso- or macro-) regionalism. See Schirm, Stefan A. (2002), *Globalization and the New Regionalism*, Cambridge: Polity Press.

25 Keating (1997:19).

26 McCrone, David (1993), 'Regionalism and Constitutional Change in Scotland', *Regional Studies*, 27(6), pp.507-512, p.508

as the attempt to safeguard regional culture, language or identity.[27] Regionalism has also been understood as a political aspiration for 'far-reaching autonomy' at the regional level, or as the advocacy of 'geographical closeness between the citizens and their elected representatives'.[28] In other contexts, regionalism has been equated with sub-national movements striving for some type of autonomy or self-government within or outside the traditional state.[29] In yet other cases, regionalism has been confused with regional consciousness, or described as something that by necessity develops within an ethnic community or from a sense of regional identity.[30] Each of these definitions may have its merits. The problem with all of them, however, is that they fail to cover the whole domain of possible regionalisms.[31] Thus, regionalism is likely to take different shapes in different time and space contexts, but regionalisms in which autonomy, identity or language play little or no role are not covered by the definitions above.

One definition of regionalism that may provide us with a more fruitful starting-point, however, is brought out by Colin H. Williams. Williams's suggestion is that regionalism is 'the attempt to optimise the interests of a region's population through the manipulation of the political process'.[32] In this conceptualization of regionalism, no presuppositions are made regarding *how* the regional population is to be understood in each case, *what* their interests are thought to be or *by what means* the political process is manipulated. Therefore, it allows us to identify different kinds of regionalisms in different time and place contexts. Williams's definition also proves to be particularly helpful if the aim is to analyse the beliefs, norms and values on which any particular regionalism rests.

---

27 Lindström, Bjarne (1996), *Regional Policy and Territorial Supremacy*, Stockholm: NordREFO, p.13

28 Nationalencyklopedin; *www.ne.se* (2005-03-08)

29 Scheinman, Lawrence (1977), 'The Interfaces of Regionalism in Western Europe: Brussels and the Peripheries', in Esman, M. J. (ed.), *Ethnic Conflict in the Western World*, Ithaca, New York: Cornell UP, p.67.

30 See Gren (2002a:20); Sasse, Gwendolyn (2002), 'Conflict-Prevention in a Transition State: The Crimean Issue in Post-Soviet Ukraine', *Nationalism and Ethnic Politics,* 8(2), pp. 1-26; Schrijver, Frans J. (2003), 'Electoral Performance of Regionalist Parties and Perspectives on Regional Identity in French Regions', Paper presented at *ECPR Joint Sessions*, Edinburgh, p.2

31 Rosing, Hans (1996), *Vetenskapens logiska grunder*, Helsingfors: Schildts förlag AB, p.87ff.

32 Williams, Colin H. (1997), 'Territory, Identity and Language', in Keating, M. and Loughlin, J. (eds.), *The Political Economy of Regionalism*, London: Frank Cass, p.114

Regionalism as Ideology

A fundamental assumption in this study is that regionalism can be understood and studied as an instance of a political ideology. When the concept of ideology is used here, it refers to a set of action-related ideas about how the world is and how it ought to be. Ideology is, in other words, understood as a belief system shared by a group of people and used to mobilize other people for action.[33] This definition should not be mistaken for one that claims to establish the meaning of 'ideology' once and for all. Instead, the definition should indicate that I dissociate myself from those who understand ideology primarily as a distortion of reality, as 'meaning in the service of power' or as a technique for sustaining certain relations of dominance.[34]

Ideologies functioning in the field of politics may very well be closely bound up with power, and some of them are certainly conducive to maintaining relations of dominance.[35] Whether or not that is the case is, however, an empirical question that I would prefer to respond to in the course of the analysis rather than at this initial stage. From my perspective ideologies are *not* restricted to dominant groups, but are found also among the dominated ones.[36] Belief systems shared by oppressed groups may communicate 'resistance and opposition' rather

---

33 van Dijk (2004:2); See also Funderburk and Thobaben (1989:1); Rejai (1995:3,9); Ball and Dagger (1999:5); Liedman, Sven-Eric (1989), 'Om Ideologier', in Liedman, S.E. and Nilsson, I. (eds.), *Om Ideologi och Ideologianalys*, Inst. för Idé- och Lärdomshistoria, Göteborgs Universitet, p.10.

34 This is not wholly uncontroversial. J. B. Thompson, for instance, denies that ideology should be seen as a neutral system of ideas, organized around a certain standpoint. In his vocabulary, ideology is rather 'meaning in the service of power' or a technique to reach a certain political goal. While refusing to see ideologies as 'packages of ideas', Thompson suggests that ideology analysis should be constituted by a study of the ways in which certain meanings aim at sustaining certain relations of dominance. As a consequence of this, Thompson rejects the idea that the concept should be stripped of its negative, critical sense. See Thompson, J.B. (1996). *Ideology and Modern Culture*, Cambridge: Polity Press, p.7-8; Thompson, J.B. (1984). *Studies in the Theory of Ideology*, Cambridge: Polity Press, p.127ff. For a plea for a 'neutral' understanding of the ideology concept, see van Dijk (2004); Paasi, Anssi (1996a), *Territories, Boundaries and Consciousness*, New York: John Wiley & Sons Inc, p.29.

35 van Dijk (2004:8).

36 van Dijk (2004:3); van Dijk, Teun A. (1997b). 'Discourse as Interaction in Society', in van Dijk, T.A. (ed.), *Discourse as Social Interaction. Discourse Studies: A Multidisciplinary Introduction*, Vol. 2., London and Thousand Oaks: SAGE, p.25-26

than the maintenance of a certain power structure, but they should nonetheless be understood as ideologies.[37] A dominant ideology, or an ideology held by a dominant group, is then a special case of ideology and not the only kind of ideology we could speak about.[38] By adopting a more neutral understanding of the concept of ideology, I do not renounce the researcher's right or duty to scrutinize ideologies critically.[39] Even among those who refuse to understand ideology as a distortion of reality, we find a long-lasting ambition to subject political ideas to criticism.[40]

Dimensions of Ideology

In several textbooks on political ideologies and comparative politics, efforts have been made to identify dimensions, categories, structures, and elements that recur in most ideologies. Larsson suggests that most ideologies contain elements such as fundamental values, a general outlook on people, ideas about societal transformation, ideas about the most important unit in society, ideas about how the economy should be organized and an image of utopian society.[41] Funderburk and Thobaben identify fewer and broader categories, and contend that most ideologies host 'core values, program, and leadership'.[42] Van Dijk has suggested that ideologies 'show a number of basic dimensions', such as criteria of membership and group access, ideas about the resources of the group and notions of the social position of the group in relation to others. Furthermore, it is argued, ideologies tend to cover ideas about 'actions and aims' and about 'norms and values' worth defending.[43] Rejai identifies five dimensions that partially correspond with those mentioned above. The dimensions that Rejai suggests recur in most ideologies are of a cognitive, affective, evaluative, programmatic, and social-base kind.[44]

Dimensions of this kind are of course not mutually exclusive, but exert an influence on and overlap each other to an extent that makes it hard to handle them as discrete entities. They are nonetheless of great methodological import,

---

37 van Dijk (2004:3); see also Rejai (1995:9) and Paasi (1996a:29).

38 van Dijk (1997b:26).

39 See Liedman (1989:10)

40 For an overview, see Beckman, Ludvig (2005), *Idéanalys – en introduktion till det kritiska studiet av politiska budskap.* Stockholm: Statsvetenskapliga Institutionen, Stockholms Universitet, p.4.

41 Larsson, Reidar (1997b), *Politiska ideologier i vår tid*, Lund: Studentlitteratur.

42 Funderburk and Thobaben (1989:1).

43 van Dijk (1997b:26).

44 Rejai (1995:4).

since they can bring a certain structure to the empirical analysis and help us identify patterns, similarities and dissimilarities in the empirical material to hand. Here, alternate readings of the empirical material and of theories on regionalism, nationalism and ideology have allowed me to identify five dimensions that prove to be useful when striving to understand the value-basis of regionalism. These dimensions, elaborated at length below, coincide largely with those identified by Mostafa Rejai in *Political Ideologies, A Comparative Approach*. The dimensions identified are, however, also informed by work done by other social scientists with a special interest in ideologies.[45] Williams's definition of regionalism has also been taken into careful consideration in the identification and elaboration of the dimensions discussed below.

World view

Several scholars contend that every political ideology hosts a world view, an 'opinion of reality'[46] or 'a view of the present'.[47] Ideologies are furthermore thought to serve explanatory ends, as they tend to provide their advocates with 'an explanation of why social, political, and economic conditions are as they are'[48] or with 'a conception of how and why the political world works'.[49] Both the descriptive and the explanatory parts of a world view can be influenced by notions of history and of the past. The past can then be represented as essentially 'good' or essentially 'bad' in comparison with present times. Past occurrences can also be cited to explain conditions today. In order to understand a political belief system, we need, accordingly, to consider its general outlook on the world. If we enhance our understanding of how the world is comprehended and explained within a certain body of ideas, we may be more capable of putting its further demands, values and policy recommendations into perspective. Here, we need to find out how the world is narrated and explained by political actors at the regional level. Do we find several contradicting world views, or is there a specific regionalist' world view to be found? How are the region's relations to the state, the nation, the European Union and the global market perceived?

---

45 See Funderburk and Thobaben (1989); Baradat (1991); Ball and Dagger (1999); van Dijk (1993, 1998, 2004).

46 Skovdahl, Bernt (1992), *Tingsten, totalitarismen och ideologierna*, Stockholm: Symposion, p. 43

47 Baradat (1991:9).

48 Ball and Dagger (1999:5).

49 Funderburk and Thobaben (1989:2).

Core Values

Political belief systems not only consist of views of the past and the present, but also communicate visions of the future and notions of how things *could* or *should* be in an ideal world.[50] Most ideologies tend to defend certain values or interests, whether these are defined in terms of freedom, equality or security. This goes well with the suggestion that regionalism involves an attempt to safeguard the perceived *interest* of the population. What these interests would be in each case (or how they are interpreted by regional representatives) is, however, an empirical question. In one space and time context, the 'right to roots' may be interpreted as an interest of the population, whereas in another such context 'economic autonomy' might be understood as a more urgent interest. Irrespective of their content, however, political statements about 'the general interest of the population' are always informed by the politician's own interests and convictions. Therefore, regional policy documents tell us little about the interests of the population in the region, but more about the values and interests of those who drafted the documents. The intention here is thus to ask how political actors in the regions narrate the 'interest of the population'. What core values are communicated, and what interests are brought out as the most important ones to defend?[51]

Strategy and Organisational Ideals

The conceptualization of regionalism employed here maintains that regionalism involves a manipulation of the political process.[52] This corresponds fairly well with the widely established notion that ideologies are *action-oriented*.[53] Hence, ideologies not only host a set of values but link these values to action in a rather explicit way.[54] The programmatic or action-related dimension of ideology tends quite naturally to take somewhat different shapes within different ideologies. Some ideologies advocate reformist plans of action whereas others are more revolutionary. Given this, we may elucidate what organizational ideals and what

---

50 van Dijk, Teun A. (1998) *Ideology – a Multidisciplinary Approach,* London: SAGE, p.69; Funderburk and Thobaben (1989:1-2); Baradat (1991:9).

51 Rejai (1995:6-9).

52 Williams (1997:114)

53 Funderburk and Thobaben (1989:4).

54 Ball and Dagger (1999:7); Rejai (1995:9); van Dijk (1998:70); Funderburk and Thobaben (1989:4); Baradat (1991:9); Skovdahl (1992:43).

political methods a regionalist ideology would advocate.[55] We may ask how classical issues such as participation, division of power and the role of leaders are conceptualized within the documents analysed.[56] We may also investigate what role organizational boundaries and hierarchies are thought to play in political organisation, and by what means regional representatives aim to manipulate the political process.

Means of Mobilization

It is widely acknowledged that political ideologies appeal not only to people's reason but also to their emotions. Rejai contends in this matter that ideology hosts 'elements of emotionality', while Baradat asserts that ideologies tend to be 'stated and presented in motivational terms'.[57] Funderburk and Thobaben stress that ideologies are far from being 'entirely rational or logical systems of thought', but that 'faith and emotional commitment are key ingredients as well'.[58] It has also been observed that advocates for different ideologies tend to romanticize different things. Some may embellish 'the family', 'the nation' or what they perceive as 'the race', whereas others proclaim the moral superiority of the working class or the nobility of a certain life style. Some political slogans tend to stress a glorious past, whereas others communicate social progress and the bright prospects of the future.[59] In view of that, we need to analyse the mobilizing aspects of regionalism. Studies of nationalism and, to some extent, of regionalism, indicate that feelings of belonging, shared ancestry and attachment to what is perceived as one's own 'cultural space' are often used to mobilize public support. Since the regions examined here are administrative constructs rather than ethnic communities, there may not necessarily exist grounds for claims about cultural affinity within the borders of these regions. Yet previous studies on nation- and region-building indicate that arguments about cultural homogeneity and affinity can be found also in 'states without nations' and in other politically defined territorial units.

---

55 For a discussion on organizational *values*, see Hansen, Bernhard and Saglie, Jo (2005), 'Who Should Govern Political Parties? Organizational Values in Norwegian and Danish Political Parties', *Scandinavian Political Studies*, 28(1), pp.1-24.

56 Hansen and Saglie (2005); Funderburk and Thobaben (1989:4)

57 Rejai (1995:6-7); Baradat (1991:1,9).

58 Funderburk and Thobaben (1989:4).

59 Funderburk and Thobaben (1989:5).

Target Group

It has been stressed that 'ideology, to be ideology, must have a mass base'.[60] In other contexts, however, it is argued that ideologies do not need such a base, but that they nevertheless are '*oriented* toward the masses'.[61] Here, the central question is not whether there is a widely mobilized regionalism in the regions studied. The central question is instead *to whom* the regionalist message is directed. In the definition of regionalism adopted here, we saw that the interest of 'the population' was given a central position.[62] Accordingly, it is important to ask how this 'population' is defined in each case. 'Population' may indeed include all those inhabiting a certain territory. However, studies of nationalism – another belief system in which 'the population' or 'the people' take centre stage – indicate that this category often turns out to be far more restricted in practice.[63] We need, therefore, to ask which groups are defined as assets and which as liabilities. We need to elucidate which groups are promoted, and which groups the regionalist policy is thought to benefit.[64] We also need to discuss the excluding 'implications of particular ways of defining a people'.[65]

TERRITORIAL IDEOLOGY

The interest shown in ideology by geographers may have been limited. This at least is how Williams argues when he contends that geographers have been more occupied with studying 'territory, symbolic places, boundary and frontiers' than with studying ideology.[66] Still, it should be of interest – particularly from a geographer's perspective – to learn more about ideologies in which geographical concepts such as space, place and territory are believed to play a prominent role.

60 Rejai (1995:9).
61 Baradat (1991:1), emphasis added.
62 Williams (1997).
63 Billig, Michael (1995), *Banal Nationalism*, London: SAGE; Painter, Joe (2000), 'Multi-level Citizenship, Identity and Regions in Contemporary in Europe', in Anderson, J (ed.) *Transnational Democracy,* London: Routledge.
64 Rejai (1995:10); van Dijk (1998:69); Ball and Dagger (1999:6).
65 Murphy, A.B. (2002), 'National Claims to Territory in the Modern State System: Geographical Considerations', *Geopolitics*, 7(2), pp.193-214, p.197.
66 Williams (1997:114).

Regionalism should then be of particular interest, since this is what Paasi has called a 'territorial ideology'.[67]

Paasi's description of regionalism as a territorial ideology is not followed by any substantial discussion of its particular characteristics. Neither are the possible differences between territorial and 'non-territorial' ideologies subject to any examination. This is unfortunate, as the concept of territorial ideologies is potentially both informative and useful if further specified. However, calls for additional studies in this field have been made before, sometimes with the argument that 'there are major gaps in our understanding of the theoretical and practical significance of territorial ideologies'.[68]

Here, it is presumed that most ideologies comprise some territorial aspects.[69] It is, however, also conjectured that in a territorial ideology, territorial aspects will play a more central role. Thus, the dimensions referred to above may qualify as fundamental ideological categories, but their *content* certainly differs across ideologies.[70] What role spatial issues such as borders and boundaries, localities and territories really play in different kinds of regionalism is, however, an open question that must be answered by means of empirical studies.[71]

The Regional Space

In analysing the value-basis of regionalism, we need to consider the concept of region and its various applications. Since the focus in this study is on two political regions – Swedish Norrbotten and German Mecklenburg-Western Pomerania – a working definition of a region could be the 'sub-national administrative unit' or the 'intermediate political level between the state and the locality'.[72] The elusiveness of the concept, however, makes such definitions highly unsatisfactory. Regions can take 'different forms in different places', and they always exist

---

67 Paasi (1996a:18).

68 Murphy (2002:194).

69 Mottos such as 'workers in all countries unite', 'socialism in one state', 'free trade' or 'preserve, protect and defend existing social order' all have territorial implications.

70 van Dijk (2004:16).

71 For a discussion of geo-political visions, see Dijkink, Gertjan (1996), *National Identity and Geopolitical Visions*, London: Routledge, p.11. See also Murphy (2002:194) and Ek, Richard (2003), *Öresundsregion - bli till!*, Lund: Lunds Universitet, p. 96.

72 MacLeod (2001:805); Keating (1998:9).

in relation to some particular criteria.[73] Apart from formal political or administrative regions, we find *functional* regions, *economic* regions, and regions that are distinguished from the surrounding society by what are perceived as particular *cultural*, *historical* or *linguistic* features.[74] Without any kind of prefix, the region concept could actually be used to describe all kinds of spatial entities. Still, we find politicians as well as academics referring to 'regions' without providing any further qualification of the concept.

The value of conceptual clarity stands out clearly in studies of nations and nationalism. Thus, the possibility of making a conceptual distinction between the nation and the state has been of central importance in this field of research. Academics may debate the advantages of some definitions over others, but nowadays there is at least a readiness to distinguish the state from the nation at a conceptual level.[75] The state is – with some variations – understood as a legal and political organization that '(successfully) claims the monopoly of the legitimate use of physical force within a given territory'.[76] The nation, in contrast, has been described as an 'imagined community', or as a 'community of people ... bound together by a sense of ... national consciousness'.[77] In studies of nationalism, congruence between the nation and the state has furthermore been identified as a fundamental nationalist ideal. Given that, the conceptual distinction between the state and the nation proves to be of great analytical value.

The importance of this distinction manifests itself also when discussing processes of state and nation building. State building has been understood as the 'creation of new governmental institutions and the strengthening of existing

---

73 Keating (1997:17; 1998); MacLeod (2001:811); Allen, J., Massey, D., Cochrane, A. (1998), *Rethinking the Region*, London and New York: Routledge, p.1

74 Keating and Loughlin (1997:2,3).

75 For another interpretation, see Connor, Walker (2004), 'Nationalism and Political Illegitimacy', in Conversi, Daniele (ed.), *Ethnonationalism in the Contemporary World: Walker Connor and the Study of Nationalism*, London: Routledge, pp.24-47. See also Keating, Michael and McGarry, John (eds.) (2001), *Minority Nationalism and the Changing International Order*, Oxford: Oxford University Press, p.1.

76 Weber, Max (1921), 'Politik als Beruf', in *Gesammelte Politische Schriften*, München, pp. 396-450. For some more recent contributions, see Seton-Watson, Hugh (1977), *Nations and States. An Enquiry into the Origins of Nations and the Politics of Nationalism*, London: Westview Press, p.1 ; Guibernau, Montserrat (1999), *Nations Without States. Political Communities in a Global Age*, Cambridge: Polity Press, p.13.

77 Anderson, Benedict (1991), *Imagined Communities,* London: Verso; Seton-Watson (1977:1); see also Hobsbawm, E.J. (1998), *Nationer och nationalism*, Stockholm: Ordfronts Förlag; Gellner (1987).

ones'.[78] The concept of nation building refers rather to the 'historical construction of "peoplehood"'[79] or to the processes through which feelings of national solidarity and belonging are constructed and reconstructed.[80] State and nation building have indeed gone hand in hand over history. Much of the 'legitimacy of state governments' has by and large been 'tied to the ability of state leaders to speak and act on behalf of nations'.[81] Still, we should not understate the value of the analytical distinction between the state and the nation. Thus, there are a number of contexts in which the vagueness of the region concept causes problems in regional studies.

Region Building

While the willingness to see the nation as an institutionalized, imagined community has increased among scholars, the region is more often understood as a genuine community and as a coherent sphere of interests, functions and identities.[82] 'The region' is not seen as the result of a political process to the same extent as 'the nation', but is more often described in terms of folklore and 'genuine' cultural roots.[83] There is indeed a growing willingness within the social sciences to understand regions as socially constructed categories, but the constructionist perspective still seems to be weaker in regional studies than in the study of nations and nation building.

Here, one of the basic assumptions is that regions quite generally are the result of region-building processes.[84] Due to the vagueness of the region concept, however, processes of region building can take very different shapes and aim to construct very different things. Region building can in one context refer to some very explicit processes. This applies, for instance, when new political constructs

---

78 Fukuyama, Francis (2004), 'The Imperative of State-building', *Journal of Democracy*, 15(2), pp.17-31,

79 Murphy (2002:194).

80 Johansson, Rune (1999), 'Ett odödligt sinnelag', in Olsson, Erik (ed.), *Etnicitetens Gränser och Mångfald*, Stockholm: Carlssons Bokförlag, pp. 285-353,

81 Murphy (2002:195).

82 Gidlund and Sörlin (1993:171).

83 See Gren (1999:75ff); Törnqvist (1993:22). For a discussion on this, see Aronsson, Peter (1995), *Regionernas roll i Sveriges historia*, Stockholm: Fritzes, pp.205f.

84 For a similar discussion, Paasi, Anssi (1996b), 'Regions as Social and Cultural Constructs: Reflections on Recent Geographical Debates', in Idvall, M, and Salomonsson, A. (eds.), *Att skapa en region – om identitet och territorium*, Stockholm: NordREFO, p.90-107; Allen *et al.* (1998:1).

such as cross-border regions or new sub-national units are 'invented' and provided with their own formal institutions.[85] These processes can be managed from above, that is, by a state intending to establish new territorial units, but they can also be managed from below. In other contexts, region-building efforts are of a more subtle kind and aim at fostering feelings of regional belonging or at providing an existing political construct with additional meaning. Attempts to foster feelings of regional belonging may be more widespread at the regional level, yet there are examples of how state authorities, too, have encouraged regional identities.[86]

The distinction between nation and state building set out above merges to some extent with the one Brubaker applies when he distinguishes polity-seeking nationalist movements from 'nationalizing states'. Polity-seeking nationalist movements aim, according to Brubaker, to 'establish or up-grade an autonomous national polity'. Nationalizing states are then defined as formally existing states that are 'conceived by their dominant elites as … insufficiently "national" in a variety of senses'.[87] In these nationalizing states, we find the elite making a variety of efforts for nationalizing the territory of the state and for making the people more aware of their nationhood. Thus, in a state-centric system, state governments could – with more or less success – use and control resources in order to make the 'bounded territory of the state the primary focus of economic, political and cultural identifications for its citizens'.[88]

While employing Brubaker's classification at the regional level, we may distinguish between regionalist movements aiming at providing 'their' 'cultural territory' with its own public institutions[89], and 'community-building regions' in which the regional public institutions are already established. There are indeed a number of examples where these two kinds of region building merge into a

---

85 See Veggeland, Noralv (1994), *Regionbygging over genser i det nye Europa – Aktörer, areaer or regimer*, Stockholm: NordREFO, p.15; Maskell, Peter and Törnqvist, Gunnar (1999), *Building a Cross-Border Learning Region*, Copenhagen: Copenhagen Business School Press; Jerneck, Magnus and Gidlund, Janerik (2001), 'Region-building, Democracy and Changing Policy-Styles in Sweden', Paper presented at *Arena, Center for European Studies*, Oslo.

86 See Hobsbawm, E and Ranger, T. (eds.) (1983), *The Invention of Tradition*, Cambridge: Cambridge University Press.

87 Brubaker, Rogers (1997) *Nationalism Reframed. Nationhood and the National Question in New Europe*. Cambridge: Cambridge University Press, p.79.

88 Anderson, J. and O'Dowd, L. (1999), 'Borders, Border Regions and Territoriality: Contradictory Meanings, Changing Significance', *Regional Studies*, 33(7), pp.593-604, p.598

89 Lindström (1996:11ff); see also Jerneck and Gidlund (2001).

multifaceted region-building process.[90] In these cases, it would be even more valuable to have the conceptual tools needed for dissecting such processes. As this study concentrates on existing, politically defined regions, it is the latter form of region building that is of greater interest. The ambition in these 'community building regions' may be to form a 'territorial frame of reference',[91] to support regional identities or, at least, to foster a sense of regional belonging or consciousness. The region building efforts then aim at transforming a political or administrative geo-political entities into culturally coherent spaces, bound together by a sense of regional consciousness.[92] Still, the difference between region building aiming at building institutions and region building aiming at building communities should not be overstated. Region building in all its forms must be considered as associated with attempts to build institutions *and* with attempts to mobilize feelings of regional solidarity and belonging to these institutions.[93]

Borders, Boundaries and Territories

The attempts in community-building regions to construct 'peoplehood' or to promote feelings of regional belonging are 'almost always pursued in the context of a territorial ambition'.[94] These territorial ambitions can indeed turn out differently, but frequently they involve attempts to claim the coherence of a particular territory.[95] In view of this, we need to discuss the role of borders and boundaries in regionalism.[96]

Parts of the political borders of both Norrbotten and MV coincide with the outer state borders of Sweden and Germany respectively. Certainly, the meaning and significance of borders can 'vary dramatically over space and time'.[97] In

---

90 See Ek (2003); Jensen, Ole B. and Richardsson, Tim (2004), *Making European Space. Mobility, power and territorial identity*, London and New York: Routledge.

91 Keating, Michael and Loughlin, John (2002), 'Territorial Policy Communities and Devolution in the United Kingdom', *EUI Working Papers, 2002/1*, Florence: European University Institute.

92 Löfgren Orvar (1993b) 'Det nationella i det globala' i Hemer, O. (ed.) *Kulturen i den globala byn*, Lund: AEGIS Förlag

93 For some examples, see Johansson, Rune (1984), 'Varieties of Conflict Development', in Tägil, S. (ed.), *Regions in Upheaval*, Solna: Esselte Studium, pp.44-94.

94 Murphy (2002:196).

95 Murphy (2002:193).

96 Anderson and O'Dowd (1999:598).

97 Anderson and O'Dowd (1999:594); see also Keating, (2004:378).

some cases, the presence of a state border has severe implications for political mobilization or for peoples' everyday lives.[98] In other cases, formal boundaries are referred to as being 'invisible' and consequently considered less influential.[99] We may not presume that the guarding of the formal, outer border of the region is a central issue in any of the regions studied. This is, however, not to say that these borders are essentially unproblematic. Political borders can be widely accepted and remain undisputed in formal terms, but still be challenged by cultural, functional and other boundaries.[100] Borders can furthermore be used for mobilizing and motivational ends, also when they appear to be very uncontroversial.

In human geography, borders are commonly referred to as being more strict and less negotiable than the softer, more 'porous', elusive and changeable boundaries.[101] The meaning and significance of a formal, political border can certainly vary greatly between different groups in society, but the mere existence of it is, however, more a fact than a contention. Boundaries are, to the contrary, matters of subjective judgement and belief. Thus, boundaries can be found within a distinct spatial entity; they can transgress the formal borders of that entity and they can incorporate it within a larger one. Unlike with formal state borders, the existence of boundaries is largely a question of interpretation. Accordingly, we need to explore the ways in which boundaries are represented and reproduced in the regions studied.[102] What boundaries are highlighted and what boundaries are 'ignored, taken-for-granted and seen as peripheral, not just in the literal geographical sense but in political and social terms'?[103]

---

98 Cyprus, Palestine/Israel and the 'post-war division of Europe by the "Iron Curtain"' are mentioned as such examples by Anderson, J., O'Dowd, L. and Wilson, T.M. (2002). 'Why Study Borders Now?', *Regional & Federal Studies*, 12(4), pp.1-12.

99 See Van Houtum, Henk and Strüver, Anke (2002), 'Borders, Strangers, Doors and Bridges', *Space & Polity*, 6(2), pp. 141–146; Tronvoll, Kjetil (1999), 'Borders of Violence – Boundaries of Identity: Demarcating the Eritrean Nation-state', *Ethnic and Racial Studies*, 22(6), pp.1037-1060, p.1041.

100 Anderson and O'Dowd (1999:595).

101 For a discussion of boundaries as bridges and as barriers, see Petersson, Bo and Clark, Eric (eds.) (2003), *Identity Dynamics and the Construction of Boundaries*, Lund: Nordic Academic Press, p.12f. For a discussion on definitions of borders and boundaries respectively, see Tronvoll (1999).

102 Anderson and O'Dowd (1999:595ff); see also Paasi, Anssi (2000), 'Re-constructing Regions and Regional Identity', *Lecture Manuscript, Nijmegen*, Netherlands, p.3.

103 Anderson, O'Dowd and Wilson (2002:3).

Who are the Region Builders?

If regions, regional boundaries and feelings of regional belonging are results of region-building processes, we must ask who the region builders are. Hence, regions are not constructs of their own making but are created, institutionalized and filled with meaning by actors within and outside the particular region.[104] Regions are therefore to be considered as political projects created by elites or, at least, by 'different organised interests'.[105] The existence of regions is, as Neumann has argued, 'preceded by the existence of region-builders'.[106] These region-builders are described as political actors who 'as part of some political project, imagine a spatial and chronological identity for a region, and disseminate this imagined identity to others'.[107]

The argument that regions are preceded by region-building efforts is in a way a 'stretching'[108] of Gellner's well-known argument that 'it is nationalism which engenders nations, and not the other way around'.[109] Many scholars acknowledge the importance of elites in the construction of nations. Murphy contends in a discussion on this matter that '[p]olitical leaders of all states' have been 'engaged in an effort to construct a unified sense of nation'.[110] Opinions differ over the reasons for these leaders' success or failure, but the rise of modern nationalism in Europe is commonly traced to a rather small section of the population.[111]

---

104 Süssner, Josefina (2002), 'Culture, Identity and the Region', *Informationen zur Raumentwicklung*, Heft 4/5, Bonn: Bundesamt für Bauwesen und Raumordnung, pp.199-206.

105 Österberg, Eva (1994), 'Tradition och Konstruktion – Svenska regioner i ett historiskt perspektiv' in, Blomberg, B. and Lindquist, S.O. (eds.), *Den Regionala Särarten*, Lund: Studentlitteratur, p.29.

106 Neumann, Iver B. (1994), 'A Region-Building Approach to Northern Europe', *Review of International Studies*, 20(1), pp.53-74, p.58

107 Neumann (1994:58).

108 For a discussion about conceptual stretching, see Sartori, Giovanni (1970). 'Concept Misformation in Comparative Politics', *American Political Science Review*, 64(4), pp.1033-53.

109 Gellner, Ernest (1983), *Nations and Nationalism*, Oxford: Blackwell, p.55. For a similar argument, see Keating (1998:87).

110 Murphy (2002:197).

111 See Breuilly, John (1994), *Nationalism and the State*, Manchester: Manchester University Press; Hobsbawm, E.J (1990), *Nationas and Nationalism since 1780*, Cambridge: Cambridge University Press. For a slightly different approach, see Smith, A.D (1995), 'Gastronomy or Geology? The Role of Nationalism in the Reconstruction of Nations', *Nations and Nationalism*, 1(1), pp. 3-23.

Accordingly, it is 'imperative that we seek to understand the ... territorial ideologies of groups that are in power'.[112]

Given the sound arguments for examining the role of elites in processes of nation or region building, one may ask if there is any elite at all to talk about in peripheral regions.[113] Hence, periphery is a concept that implicates 'powerlessness in relation to the centre, whereas *elite* indicate the presence of power'.[114] Nevertheless, some actors are more capable than others of producing 'change in the regional political landscape', and these actors may be regarded as elites or as potential region builders.[115]

VARIANTS OF REGIONALISM

So far, it has been suggested that regionalism can be studied as an ideology, and that territorial aspects may play a part in all dimensions of this ideology. Thus, even if one can observe what van Dijk has called 'fundamental ideological categories', we may expect that these categories will be given different content in different ideologies.[116] In line with this, it can be assumed that regionalism – like any other ideology – can take different forms and host different demands in different time and place contexts. Studies of the development in western Europe since the late 1960s do provide evidence that regionalism and region building take different forms over time and place, though there are few systematic studies of the value-basis of different forms of regionalism.

Old regionalisms

As indicated in the introduction, it is commonplace to distinguish between 'old' and 'new' regionalism. Accounts of these forms of regionalism tend to be quite

---

112 Murphy (2002:196f).

113 Gidlund J., Sörlin S., Gidlund S. (2000), *Ensam hemma - Den norrländska elitens nya syn på regional utveckling*, Umeå: CERUM, Umeå University, p.9ff.

114 Gidlund *et al.* (2000:9ff).

115 Christopoulos, Dimitrios (2001), 'Regional Elites and Public Entrepreneurship', Paper presented at *ECPR Joint Sessions of Workshops*, Grenoble. For a discussion on 'centers in the periphery', see Galtung, J. (1980), 'A Structural Theory of Imperialism', in I. Vogeler and A.R. de Souza (eds), *Dialectics of Third World Development*. New Jersey: Allanheld, Osmun & Co., pp. 261–298.

116 van Dijk (2004:16).

static. Old regionalism is then used a generic term for all kinds of regionalist, separatist or sub-nationalist movements that were on the political agenda in the 1960s and 70s. Furthermore, accounts of 'old regionalism' tend to revolve around political events in peripheral, less favoured regions or around 'ethnic' or 'cultural' regions with sub-nationalist demands.[117]

A common feature in accounts of old forms of regionalism is the alleged experience of *threat*. Thus, in some cases a particular form of defensive regionalism is said to have appeared in regions where central sectors of the economy were understood as being 'threatened by modernization or European competition'.[118] In other cases, parts of the regional mobilization in western Europe during the 1970s is said to have merged with grass-roots protests against the threat of large-scale environmental pollution.[119] In yet other cases, regional minority languages and cultures were considered as being threatened by a homogenizing central state and by the modernist idea that regionally distinctive cultural traditions would obstruct economic growth and development within the nation-state.[120] Regional movements in those days, however, also accommodated pressures for 'democratization and participation' and opposition against the centralizing state.[121] Thus, the state and its alleged efforts to homogenize and centralize society were seen not only as a threat to the economy or to regional minority cultures. State centrism was also thought to cause a democratic deficit and, later, to have resulted in a 'crisis of territorial representation'.[122] Older forms of regionalism are thus said to have emerged and existed 'in relation to the state to which their demands were addressed'.[123] There are indeed examples of regional or sub-state nationalist movements establishing international relations

---

117 Hechter (1975); Harvie, Christopher (1994), *The Rise of Regional Europe*, London: Routledge, p.4, 54 Gren (2002b:81). For an overview of voices questioning this interpretation, see Olzak, (1998).

118 Keating (1997:23); See also Gidlund and Sörlin (1993:162).

119 Bort, Eberhard (2000), 'European Borders in Transition: The Internal and External Frontiers of the European Union', in *Boundary Permeability in Perspective, Papers Submitted to the Canadian Centre for Foreign Policy Development Conference*, University of Durham., p.65.

120 Lipset, Seymour (1981), 'The Revolt Against Modernity', in Torsvik P. (ed.), *Mobilisation, Center-Periphery-Structure and Nation-Building*, Bergen: Univ.Forl., pp.451-500, p.472.; Schmitt-Egner (2002:188). For a discussion on this, see also Österberg (1994); Keating (1997:23).

121 Keating (1997:23).

122 Keating (1997:24).

123 Keating (1997:24).

and taking foreign action at the very beginning of the 20th century.[124] In general, however, the international context played a relatively minor role for regional political action.[125]

New Regionalisms

By the late 1980s and early 1990s, globalization and European integration provided regionalism with an extended context, and regional demands were no longer addressed exclusively to the central state.[126] We could observe what Burch and Gomez have called an 'increasing engagement of sub-national governments on EU related issues'. Sub-national actors were to a greater extent than before involved in 'Structural Funds programming' and in 'direct sub-national lobbying activities in Brussels'.[127] Regional actors now started to address policy makers at the EU level with their demands and they became increasingly involved in 'para-diplomacy', in structures of multi-level governance and in different kinds of trans-regional cooperation.[128] Thus, regions were no longer seen as being fully politically dependent on their respective states, and instead were more commonly referred to as 'actors in their own right'.[129]

The political strategies referred to above emerged simultaneously with the advance and relaunch of the European Community. Thus, ever since the 1980s the European Community had expressed an increasing ambition to strengthen the cooperation between the Community, the state and the regional level. The extension of Structural Funding in the late 1980s and the decision to adopt the principle of partnership (which gave regions an explicit role in structural funding

---

124 For one such example, see Ugalde Zubiri, Alexander (1999). 'The International Relations of Basque Nationalism and the first Basque Autonomous Government (1890-1939)', in Aldecoa, F. and Keating, M. (eds.). *Paradiplomacy in Action: The Foreign Relations of Subnational Governments*, London: Frank Cass, p.170.

125 Keating (1997:25); Gren (1999:15).

126 Cole and Loughlin (2003:266); Keating (1997:25).

127 Burch, Martin and Gomez, Ricardo (2002), 'The English Regions and the European Union', *Regional Studies*, 36(7), pp.767-778, p.768.

128 Aldecoa, F. and Keating, M. (eds.) (1999), *Paradiplomacy in Action. The foreign relations of Subnational Governments*, London: Frank Cass; Krämer, Raimund (ed.) (1998), *Regionen in der Europäischen Union, Beiträge zur Debatte*, Berlin: Potsdamer Textbücher.

129 Cole and Loughlin (2003:266); see also Jerneck, Magnus (2003), 'Regional Lobbying och Flernivåstyrning', in Tallberg, P. (ed.), *Regionernas Europa*, Region Skåne och Västra Götalandsregionen, pp.82-85.

processes) are said to have increased the importance of decision-making at the regional level.[130] The adoption of the Maastricht Treaty, the principle of subsidiarity and the establishment of the Committee of Regions (CoR) have also been central elements in this process.

Another prominent feature of the new regionalism is that it emerged in what Harvie has called 'bourgeois' regions, that is, central regions with significant economic and political resources.[131] According to Guibernau, the development of the Single Market and the 'rising significance of European regional policy' encouraged the emergence of 'a new kind of innovative, specialized economic region oriented towards the global economy'.[132] Regions with a strong economy now merged and adopted cross-regional cooperation for gaining an increased influence in Brussels.[133] Accordingly, many accounts of the new regionalism proceed from functional regions or city-regions equipped with enough political and financial resources to have a say in the European policy process.[134] Yet the so-called *Four Motors of Europe* – Catalonia (Spain), Lombardy (Italy), Rhône-Alpes (France) and Baden-Württemberg (Germany) – have also been mentioned as examples of regions in which the new regionalism first emerged. Growth regions such as Wales, the Cambridge region, Nord-Pas-de-Calais and the Oresund region have also been used to illustrate the emergence of a new kind of regionalism.[135] The demands from these regions on the EU were not about 'rights to roots' or about the protection of threatened economic sectors. Rather, the demands concerned 'less state intervention in the economy, reduced public spending (and) cuts to redistributive regional economic development'.[136]

New regionalism is, however, not only understood as a contemporary political tendency in western European 'core regions'[137]. The concept has also been used

---

130 The partnership principle had indeed been integrated in the regional policy ever since the European Regional Development Fund (ERDF) came into existence, but it was not until the 1988 reform that the role of sub-national actors was institutionalized.

131 Harvie (1994); see also Gren (1999:15); Keating (1997:24); Törnqvist (1993:42).

132 Guibernau (1999:173).

133 Harvie (1994:3); Lash, Scott and Urry, John (2002), *Economies of Sign and Space*, London: SAGE, p.282.

134 MacLeod (2001:808).

135 Gren (1999); Keating (1998); Lindeborg, Lisbeth (2003), 'Framgångsrika Regioner' in Tallberg, P. (ed.), *Regionernas Europa*, Region Skåne och Västra Götalandsregionen, p.67.

136 Rumford (2002:178).

137 Harvie (1994:65).

for describing an emerging theory for regional economic development.[138] European integration, globalization and the alleged changing capability of the state are said to have opened up a new way of thinking about regional economic activities. Some have described this as a 'new regional development paradigm'[139], whereas others have called it an 'institutional turn' in economic theory.[140] The new regionalist perspective on economic development is introducing the notion that regional development should not be achieved 'from above', by government subsidies or through special grants for the periphery. Instead, regional development should grow 'from below' and by way of the efforts of the region itself.[141] In this tradition regional variations in growth and development are no longer explained primarily by reference to structural conditions such as location, the standard of the infrastructure or differences in access to fixed assets.[142] Instead, it is 'soft factors' or 'endogenous potentials' such as 'habituated routines and conventions',[143] cultures and identities, the level of trust and the condition of the civil society that are understood to constitute and order economic life, and to be decisive for economic growth and development.[144] Also, when new regionalism is perceived as a theory of economic development, the focus has been on growth regions and regions known for their innovative capacity. Yet many growth coalitions from less favoured cities and regions have, as MacLeod points out, been influenced by what is understood as 'intelligent spaces' and exemplary growth regions.[145]

---

138 MacLeod (2001:805).

139 Keating *et al.* (2003:14).

140 Amin, Ash (2001), 'Moving On: Institutionalism in Economic Geography', *Environment and Planning A*, 33(1), pp.1237-1241, p.1237; Tomaney, John and Ward, Neil (2000), 'England and the "New Regionalism"', *Regional Studies*, 34(5), pp. 471-478.

141 For a discussion on this, see Lovering (1999); Jones, Martin and MacLeod, Gordon (1999), 'Towards a Regional Renaissance? Reconfiguring and Rescaling England's Economic Governance', *Transactions of the Institute of British Geographers*, 24(3), pp.295-313; Hudson, Christine (2005), 'Regional Development Partnerships in Sweden: Putting the Government Back in Governance?', *Regional & Federal Studies*, 15(3), pp. 311-327, p.313ff.

142 MacLeod (2001:807).

143 Amin (2001:1237).

144 Ceccato and Persson (1999:19f); Amin (2001:1237).

145 MacLeod (2001:809).

Regionalisms – similarities and differences

To treat old and new regionalism as distinct or separate entities is certainly not without problems. Old regionalism and new regionalism constitute, as indicated above, ideal types in which considerable variations can be found. Regionalism is still today 'associated with backwardness and opposition' in some parts of Europe, whereas it is associated with progress and modernity in others.[146] Besides, regionalism in all its forms must be understood as being underpinned by several ideological doctrines.[147] Yet the above exposition indicates that in both old and new regionalism there is an ambition to increase the economic, political or cultural importance of a particular territory. Accordingly, both forms of regionalism are essentially spatial in nature.

Nonetheless, there are differences between older and contemporary forms of regionalism that are worth noting. Recalling the argument that regionalism amounts to 'the attempt to optimise the interests of a region's population through the manipulation of the political process', we can identify some central differences between old and new types of regionalism.[148] *First*, the 'interest' in old regionalism more often concerned the protection of threatened economic sectors, environments, minority languages or cultures. In the new regionalism, the interest is rather identified as increased levels of economic autonomy from the state.[149] The regions in focus during the old regionalist era were to a great extent been conceived as marginal or threatened. New regionalism, by contrast, has been explicitly located in central and 'successful' western European regions. *Second*, the ways in which the political process has been manipulated have also varied over time. In the 1960s and 70s, interaction with the central state was the main political strategy, and in some cases this interaction took rather violent forms.[150] Recent forms of regionalism have, to the contrary, been characterized by an increasing eagerness to interact with other regions and with the European Union, primarily by means of lobbying and diplomacy.

The ways in which 'the regional population' is understood and described in these two types of regionalism is still a fairly open question. Recent forms of

---

146 Keating (2001:220).

147 See Breuilly (1994:12-13).

148 Williams (1997:114).

149 Rumford (2002:178).

150 Commonly mentioned examples are Northern Ireland, Corsica and the Basque Country. See Loughlin, John (2000), 'The Transformation of the State and New Territorial Politics', Paper presented at *Center for International Studies and Research*, Paris, p.8.

European regionalism have indeed been described as modernizing, inclusive and forward-looking, whereas previous forms of regionalism have been characterized as more tied up with cultural and separatist 'ethno-nationalist' ideals and claims. Old regionalism has thus been perceived as more 'bent on excluding multicultural identities by political pressure and violence', and 'ethnically' oriented versions of old regionalism have more often been conceptually connected to nationalism.[151] To what extent new regionalism really is an integrative and inclusive force is, however, still a question for empirical research.

151 Bort (2000:65); Schmitt-Egner (2002:188).

2. Analysing Regionalism: Methodological Considerations

In this study it is suggested that regionalism can be understood as an instance of a political ideology – a 'belief system' that is shared by a group of people and used to mobilize other people for action. Regionalism and other ideologies manifest themselves in many different ways in society, and can therefore be studied through a variety of methods. Here, the focus will be on the ways in which regionalism manifests itself discursively in political texts. The approach to 'discourse' adopted here is influenced by the tradition of critical discourse analysis. The term 'discourse' thus refers here to what van Dijk has called a 'communicative event' or – to use Fairclough's definition – to 'spoken or written language use'.[1] This definition implies that material structures and social practices are excluded from the concept of discourse.[2] This conceptual distinction does indeed involve a separation of elements that in fact are 'dialectically intertwined'.[3] Yet the distinction is not meant to imply that the use of language is independent of material structures or social practices. To the contrary, the use of spoken or written language is seen as a social event, very dependent on its context.[4]

Discursive Manifestations of Regionalism

The decision to study regionalism through its 'discursive manifestations'[5] does not imply a rejection of the idea that regionalism can be both affected by and manifested in practices other than the discursive ones. To the contrary, social and

1 van Dijk (1998:62-66); Fairclough, Norman (1992), *Discourse and Social Change*, Cambridge: Polity Press, p.62
2 van Dijk (1998:197).
3 Brenner, Neil (2001), 'The Limits to Scale? Methodological Reflections on Scalar Construction', *Progress in Human Geography*, 25(4), pp.591-614, p.593.
4 Fairclough, Norman (2003), *Analysing Discourse. Textual Analysis for Social Research*, London and New York: Routledge, p.21,26; Liedman (1989:12); van Dijk (1997b:31).
5 van Dijk (1997b:31-32).

material structures can be central to the construction of meaning or to the rise of political belief systems. The decision to focus on the ways in which regionalism comes out in text and talk at the regional level was, rather, taken for methodological reasons. Discourses make ideologies 'observable', as text and talk allow people to '*express* or *formulate* abstract ideological beliefs'.[6] Ideologies may be possible to 'read off' in material structures and in social practice, too, but it is largely through discourse that political ideologies are 'acquired, expressed, learned, propagated, and contested'.[7]

Another and equally important reason for studying the use of text and talk in politics is that texts and talks contributes to the production of meaning. That is to say, with a discursive approach to politics and ideology, one ceases to regard political texts as means for transporting ideas only.[8] Instead, discourses are understood to contribute to the very production and reproduction of meaning and of political thought.[9] The capacity of discourses to standardize and to order thoughts and 'knowledge'[10] makes them vital elements in political life. Hence, an analysis of the discursive manifestations of ideology informs us about agenda setting, about political priorities, and about how political actors make sense of the world. This, in turn, is important, since the way in which actors make sense of the world has consequences for how they handle it. Consequently, discursive representations are significant for social practice as well as for changes and maintenance of material structures.

More generally, the discursive approach to politics and ideology implies that we do not need to focus only on the ways in which political actors aim to *solve* political problems. A basic assumption in a discursive approach to politics and ideology is that it is 'inappropriate to see governments as responding to "problems" that exist "out there" in the community'. Instead, problems as well as solutions are understood as being socially constructed.[11]

---

6 van Dijk (2004:10); van Dijk (1998:192).

7 van Dijk (2004:11). For a discussion on problems with 'reading off' ideologies in social practice, see van Dijk (1997b:30-31).

8 For a similar approach, see Liedman (1989:31).

9 Garrison, Raymond M. and Massam, Bryan H (2001), 'Changing Governments and Changing Environmental Policies in Ontario: a discourse analysis', *Local Environment*, 6(3), pp.327–349, p.330

10 Thompson, John B., 'Introduction' in Bourdieu, Pierre (1991), *Language and Symbolic Power*, Cambridge: Polity & Basil Blackwell, p.4; Hoggart, K., Lees, L., and Davies, A. (2002), *Researching Human Geography*, London: Arnold, p.164.

11 Bacchi, Carol (2000), 'Policy as Discourse: what does it mean? Where does it get us?', *Discourses: studies in the cultural politics of education*, 21(1), pp. 45-57, p.48.

Even if I question the existence of 'problems' as such, I do not oppose the idea of a reality external to our thoughts. Thus, like many who apply critical discourse analysis, I am influenced by critical realist thought.[12] A critical realist accepts the idea that there is a 'reality out there', but recognizes that this reality is 'inevitably constructed through our interpretations, leading to the existence of multiple understandings of those "realities"'.[13] Whereas some may consider certain elements of 'reality' as being very problematic, others see no problem at all.

The critical realist approach implies that I understand discourses as *constituents* of reality. Social and material practices constitute reality too, but there is no given congruence between what we think we know about these practices and the ways in which others perceive or describe them. Political processes may be described in one way in political text and talk, but these texts and talks are not 'transparent representations' of how political life really is organized.[14] By reading documents only, we cannot draw conclusions about how political actors really operate on a daily basis. Neither are we in a position to judge the outcome of policy recommendations or the implementation of decisions. This, however, does not mean that discursive representations are less important. To the contrary, discursive representations of a region, of its problems and of its resources are important parts of reality, even if social scientists question the validity of some of the actual representations.

With this approach to discourse, my ambition is not to decide how well the discourses observed 'match' social and material practices in the regions.[15] Rather, the ambition to scrutinize these discourses critically manifests itself in a 'systematic refusal' to accept the expressions used and the causal connections maintained in the regional policy discourse.[16] In my analysis, I hope to expose both regularities and inconsistencies in the texts analysed, to problematize them and to explain them 'with reference to … a structural context'.[17]

---

12 See, for instance, Chouliaraki, Lilie and Fairclough, Norman (1999), *Discourse in Late Modernity*, Edinburgh: Edinburgh University Press.

13 Hoggart *et al.* (2002:307).

14 Atkinson, Paul and Coffey, Amanda (2004), 'Analysing Documentary Realities', in Silverman, David (ed.), *Qualitative Research. Theory, Method and Practice*, London: SAGE, p. 58.

15 Sahlin, I. (1999), 'Diskursanalys som sociologisk metod', in Sjöberg, K. (ed.), *Mer än kalla fakta. Kvalitativ forskning i praktiken*. Lund: Studentlitteratur, p.83-106.

16 Liedman (1989:30).

17 Marsh, David and Stoker, Gerry (eds.) (1995), *Theory and Methods in Political Science*, London: Macmillan Press, p.14.

THE EMPIRICAL MATERIAL

In this study the empirical material consists almost exclusively of written records. From a methodological point of view, written records have both advantages and disadvantages. One advantage is that they may be considered 'more reasoned statements than verbal language'. While analysing texts, one could presume that the authors had an 'opportunity to revise the text prior to its dissemination'.[18] Another advantage of written documents is that they allow the researcher to collect data 'beyond what is achievable through interviews or direct observation'.[19] This advantage, however, can turn into a problem if one happens to study a field in which a massive amount of texts is produced, and where there are no established guidelines for selecting sources. Thus, in a qualitative work it is not always possible to predetermine which sources are relevant for the study as such.

Selection of Sources

The selection of sources has followed some general principles. To begin with, I have searched for texts produced within what van Dijk has called a political domain.[20] Since I have a special interest in regionalism, I have paid particular attention to texts in which the region and its development are in focus. All records referred to can be classified as documentary sources,[21] published and available for the general public prior to my investigation. Most texts were also produced primarily for distribution to the public at large.[22] The size and the comprehensiveness of each document, however, vary considerably. What sort of document we deal with or what *genre* a certain document belongs to has yet been of minor importance. Thus, it is not the form or the style of the documents that is in focus, but the values and norms they are communicating.[23] The time-frame of the study stretches from 1994, when general elections were held in both Norrbotten and Mecklenburg-Western Pomerania, to the present. Most emphasis, how-

18 Garrison and Massam (2001:333).
19 Hoggart *et al.* (2002:124).
20 van Dijk (1998:196); see also van Dijk (2004:11).
21 Documentary sources could be interpreted as documents that were not seen as 'data' when they where produced. See Hoggart *et al.* (2002:120).
22 Atkinson and Coffey (2004:57).
23 Atkinson and Coffey (2004:59).

ever, is given to the two terms of office following after 1994, that is, 1994–98 and 1998–2002.

These general principles of selection have resulted in a body of empirical material consisting primarily of party programmes, regional development agendas, action plans, transcripts of speeches, and press releases. Furthermore, three sorts of regional press material have been used in the study. The first category consists of articles and columns signed by regional politicians. This category is larger in the Swedish part of the study, since the tradition of submitting so-called 'debate articles' to the press is stronger here than in the German context. The second category consists of columns signed by editorial writers. These columns have been taken into consideration, primarily because of their opinion-forming function. This category is also larger in the Swedish part of the study. The third category consists of news reports from the regional press. These reports have not been subjected to the same kind of examination as the policy documents but have been used rather for contextualizing purposes.

More generally, the collection of data has been influenced by the critical realist idea that that social scientific research is an iterative process, characterized by a continuous 'interplay of theoretical and empirical worlds'.[24] At the initial stage, the empirical material was worked out primarily in relation to the research questions presented in the Introduction. At a later stage, however, the empirical material was augmented and adjusted concurrently with the theoretical refinement. This mode of proceeding is widely acknowledged as an *abductive* approach to science. According to abductive logic, social scientific research does not consist of a series of discrete steps.[25] As an alternative, it recommends a continuous interaction between theoretical and empirical spheres throughout the research process. It recognizes that, while one works with documents, the findings tend to determine 'what you can analyse, and what you analyse structures what you look for'.[26] With this approach, it is fully accepted that the empirical material takes shape gradually rather than being predetermined at the initial stage of the work.[27]

---

24 Silbergh, David M. (2001), *Doing Dissertations in Politics – A Students Guide*, London and New York: Routledge, p.39.

25 Hoggart *et al.* (2002:156).

26 Hill, Michael R. (1993), *Archival Strategies and Techniques*, Newbury Park, Calif.: SAGE, p.6, See also Hoggart *et al.* (2002:137).

27 Alvesson, M & Sköldberg, K (1994), *Tolkning och Reflektion: vetenskapsfilosofi och kvalitativ metod*, Lund: Studentlitteratur, p.41ff; See also Hoggart *et al.* (2002:137); Marsh and Stoker (1995:15).

The texts analysed in this volume have all been produced by actors holding some kind of official mandate or function at the regional level. The documents have been produced and distributed by political parties, politicians, regional planning associations, regional authorities or different kinds of regional development coalitions. The extent to which regional representatives really have access to authoritative decision-making has been very much debated in recent years.[28] We should therefore avoid quick judgements about the actual power resources of policy makers in the two regions under study. Yet these representatives are more capable of producing 'change in the regional political landscape'[29] than those without any public political function. This implies that this is a study of elite discourse, or an analysis of an elite ideology, rather than a study of a wider social movement.[30]

The fact that this is a study of an elite ideology has some obvious repercussions on the analysis of the empirical material. Thus, given that there is a reality independent of our thinking, but different ways to conceptualize and narrate this reality, it is essential to consider the power-positions of those whose narratives we analyse. A text 'only makes sense in a context',[31] and the question of who is the producer of the text is a relevant contextual category. Thus, ideologies may be expressed and reproduced discursively, but – as van Dijk stresses – it is not discourses but actors who 'have' ideologies.[32]

Saying that actors 'have' ideologies does not imply that I reduce actors to 'value holders' in some essentialist sense. Thus, values are certainly not 'held in some internal sense', but are invoked, appealed to and – above all – acquired and formulated in cooperation with others.[33] Therefore, ideologies and discourses are collective properties rather than personal or individual ones. This notwithstanding, discourses should not be treated as an independent authority, being 'both

---

28 For some voices in this debate, see Börzel, A. Tanja (2002), *States and Regions in the European Union – Institutional Adaption in Germany and Spain*, Cambridge: Cambridge University Press; Marks, G., Haesly, R., and Mbaye, H. (2002), 'What Do Subnational Offices Think They Are Doing in Brussels?', *Regional & Federal Studies*, 12(3), pp.1 – 23; Burch and Gomez (2002).

29 Christopoulos (2001).

30 For one definition of a social movement, see Mayo, Marjorie (2000), *Cultures, Communities, Identities*, New York: Palgrave.

31 Agnew, J., Livingstone, D. and Roger, A. (1996) *Human Geography: an essential anthology*, Oxford: Blackwell, p.9 quoted in Garrison and Massam (2001:330).

32 Bacchi (2000:53); van Dijk (2004:12).

33 van Dijk (1997b:26,31).

everywhere and nowhere at the same time'.[34] Part of the power of the spoken or written word is delegated to the language from the one who speaks or writes.[35] And it is for that reason that we need to consider the power position of those whose discourses we analyse.

This provided, it must still be emphasized that an analysis of political ideas can pay more or less regard to the actors who give voice to a certain opinion. In this matter, Beckman has distinguished studies in which *ideas and arguments* are in focus from those that centre on *actors*. A study that concentrates on ideas and arguments may indeed pay some attention to the actors communicating the ideas analysed. The ambition of such a study, however, is not to determine the role of various actors in the policy process, but to advance knowledge about the ideas they communicate.[36] In the present work, it is ideas and the arguments that are the centre of attention. It is the value-basis of regionalism that is to be explained – not the actors who take part in the production and reproduction of it.

An Interpretive Reading

While doing observations we need a structure, a focus, or at least a 'preconceived idea about what it is that is important to observe'.[37] One way to structure the reading of the written texts is to identify patterns and categories that recur more frequently than others.[38] This structuring process could, of course, be informed by an inductive ideal and guided primarily by the empirical findings themselves. Van Dijk, however, suggests that we may 'avoid a rather arbitrary discovery procedure' if we 'proceed in a more systematic and theory-driven way'.[39] This procedure goes well with the abductive approach and with the

---

34 Harvey, David (1996), *Justice, Nature and the Geography of Difference*, Oxford: Blackwell, p.86-87. see also van Dijk, Teun A. (1997a), 'The Study of Discourse', in van Dijk, T. A. (ed.) *Discourse as Structure and Process. Discourse Studies: A Multidisciplinary Introduction*, Vol 1., London and Thousand Oaks: SAGE, p. 20.

35 Bourdieu, Pierre (1991), *Language and Symbolic Power,* Cambridge: Polity & Basil Blackwell, p.107.

36 Beckman (2005:7,9).

37 Hoggart *et al.* (2002:9).

38 van Dijk, Teun A (1985), 'Structures of News in the Press', in van Dijk (ed.), *Discourse and Communication*, Berlin: de Gruyter, pp.69-93; van Dijk, Teun A. (1993), *Elite Discourse and Racism*, California, London, New Delhi: SAGE, p.33; van Dijk (1997a:5); Sahlin (1999:89).

39 van Dijk (2004:14).

recommendation to pursue continuous interaction between theoretical and empirical spheres throughout the research process.

In Chapter 1 it was suggested that the value-basis of regionalism could be analysed through the consideration of five dimensions. My ambition while reading and rereading the documents referred to above has, therefore, been to examine how these five dimensions come out in the regional policy discourse. Awareness of the central role of *world views* in ideologies has made me think about the ways in which the world is explained and understood in the policy documents. Knowing that most ideologies host *core values*, I have attended to values that are represented in the texts as being superior to other values, and to values that seem to be central in the sense that other values are defined in relation to them. Furthermore, the suggestion that ideologies tend to be action-related has turned my attention towards the *organizational ideals* communicated in the regional policy discourse. Moreover, the cognizance of the *affective* dimensions to *mobilization* has motivated an analysis of the features that are romanticized and referred to in motivational and mobilizing terms. Lastly, knowledge of the import of *target groups* in ideologies has made me ask to whom the message in the regionalist policy discourse is directed, and how the regional 'population' is defined and distinguished in each case.

The five dimensions are certainly not distinct entities, but overlap and exert an ongoing influence on each other. Nonetheless, they have provided an interpretive scheme for selecting the topics on which to focus in the written documents. They have also brought a certain structure to the written account of the results of the research process. Accordingly, theories of ideology have been of outmost importance for understanding the research object as well as for systematizing the empirical findings. Apart from that, theory on regionalism and region building, nationalism and nation building has helped me to both organize and to interpret the data.[40] Geographical theory on the construction of space and place has been taken into consideration, since regionalism is fundamentally spatial in nature. As regionalism is approached as a *territorial* ideology, special attention is devoted to the ways in which borders, boundaries and territories are used and represented in policy documents from within the regions.

40 Marsh, David and Mackie, Tom (1995), 'The Comparative Method', in Marsh, D. and Stoker, G. (eds.), *Theory and Methods in Political Science*, London: Macmillan Press, p.175.

The value of having an interpretive scheme may be greater in a study where two or more cases are compared. In a comparative study it may be even more central to be clear about the objects and the properties that are to be investigated in each case.[41] Yet we need to bring out the intricacies associated with combining an analysis of policy discourses with a comparative approach. Comparative politics have often – explicitly or implicitly – operated within a rather positivist tradition, that is, a tradition in which the scientific ideal has been to find relationships between variables presumed to be consistent across time and space.[42] Owing to this, it could be argued that a comparative approach is unfeasible when analysing discourse and ideology. Political concepts, one could object, 'do not "travel" across different cultural contexts', and if they do travel, this travelling could result in fundamental differences in the meaning of the concepts across space.[43] And indeed, the meaning of a concept or an action can vary with its context.[44] Concepts do not necessarily 'apply equally in all societies', and the same phrases do not need to bear the same meaning everywhere and at all times.[45] This problem is all the more central to consider when the comparison, as in this case, covers more than one language. Thus, the question of whether 'users of different languages communicate essentially the same information' or whether language shapes the very content of a narrative, is worthy of discussion.[46]

Another difficulty with comparing the ways in which text and talk are used in politics springs from the institutional dissimilarities between the cases. In this study, the institutional differences between the two regions preclude a perfect match of the levels of observation.[47] The similarities and differences between the

---

41 Denk, Thomas (2002), *Komparativ metod – förståelse genom jämförelse*, Lund: Studentlitteratur, p.32.

42 Marsh and Mackie (1995).

43 Landmann, Todd (2003), *Issues and Methods in Comparative Politics, An Introduction*, London and New York: Routledge, p.17.

44 Landmann (2003:43-44); Hague, R., Harrop, M., Breslin, S. (1992), *Comparative Government and Politics – An Introduction*, London: Macmillan Press, p.29.

45 Landmann (2003:43-44); Peters, G.B (1998), *Comparative Politics, Theory and Methods*, Basingstoke: Macmillan Press, p.86.

46 Taub, S., Piñar, P. and Galvan, D. (n.d.), 'Comparing Spatial Information in Speech/Gesture and Signed Language', Paper presented at *International Conference on Theoretical Issues in Sign Language Research*, Gallaudet University.

47 For a discussion on observational and explanatory levels, see Landmann (2003); see also Ragin, Charles C. (1987), *The Comparative Method – Moving Beyond Qualita-*

cases will be elaborated at length in Chapter 3, but, briefly, the decision-making power of a Swedish county is much more limited than that of a German *Land*. An analysis of official documentary sources in Mecklenburg-Western Pomerania must take into consideration acts of parliament, circulars and declarations from various ministries, whereas no such documents exist in the county of Norrbotten. Here, however, the difference in legal status between the two sets of documents is of minor importance. It is the values and beliefs communicated in the documents that are to be analysed, not the legal status of the documents or of the institutions that produced them.

Understanding in Context

Here, it is assumed that discursive manifestations of ideology *can* be compared in time, in space, and across language borders, given that they are considered in relation to their overall context.[48] That is, we need to consider the situatedness of both ideology and discourse. Thus, social agents or political actors are not 'free' in the sense that the can say, think or propose anything. Instead, they are constrained by the social structure in which they work.[49] Therefore, we need to reflect not only on the power position of those who use language, but also on the general context in which they do it.[50]

This ambition brings up the question of what a context is.[51] This question will be elaborated in more detail in Chapter 3, but, briefly, I would suggest that the context is constituted by those things that function as 'background, setting, surroundings, conditions or consequences'[52] in the two regions respectively. The context of a case is drawn up in relation to theory and to the actual research question. Contextualization as well as analysis, however, is also influenced by the researchers' own thoughts and pre-knowledge of the cases. I have no personal attachment to any of the regions studied but have undertaken several field trips to Norrbotten in recent years. I spent a full academic year in Mecklenburg-

---

tive and Quantitative Strategies, Berkeley, Los Angeles, London: University of California Press, p.9.

48 Ragin, Charles C. and Hein, Jeremy (1993), 'The Comparative Study of Ethnicity', in Stanfield, J.H and Dennis, R.M (eds.), *Race and ethnicity in research methods*, Newbury Park, CA: SAGE, p.270; Landmann (2003:43-44).

49 For a discussion on social structures, see Fairclough (2003:22ff). See also van Dijk (1997a:19; 1997b:11).

50 van Dijk (1997a:2).

51 See Marsh and Stoker (1995:14).

52 van Dijk (1997b:11).

Western Pomerania and thereby have had the opportunity to deepen my understanding of the cases. I had the opportunity to attend political meetings and seminars, and I have talked to political representatives and to individuals in both regions. These talks took the form of conversations rather than proper interviews, but have nonetheless been of great value. The experience of 'having been there' is not made visible through references to my own observations or through quotations from personal conversations. The time spent in the field has nevertheless provided me with a deeper understanding of social and political life in the two regions. I have also learned that the different languages – Swedish and German, translated into English – form a central part of the context. It has become evident that there are words and sayings that are so intertwined in their cultural and political context that they would lose much of their meaning if they were translated literally. In cases where much of the meaning disappears with translation, I have kept the original word or phrase and explained its meaning in a footnote.

Making Use of Methodology

At this point, it should be clear what my empirical material is, how it has been selected, and how I intend to analyse it. It should also be clear that I approach regionalism as an instance of a political ideology that can be analysed through a consideration of the five dimensions set out above. The main merit of employing this model is that it allows us to analyse a value-basis in a systematic way. The five dimensions provided me with a rough structure in a late phase of the empirical observation, in the analysis and particularly in the written account of the research findings.[53]

It has also been stated that theory has been used throughout the research process. Initially, theory on both ideology and regionalism was used as a means to generate relevant questions and as a guide for what to investigate. Theory thus provided an ability to 'foreground' certain aspects of the world, and made it possible to 'focus on particular aspects of reality'.[54] At a later stage, theories of regionalism, nationalism and ideology were central for structuring the observations and the analysis, and for contextualizing the empirical findings. Lastly, the written account of the research process came to be very much guided by theories on ideology. Thus, unlike with many comparative studies, I do not intend to press the cases into two separate chapters and then analyse them separately.

53 Marsh and Stoker (1995:17)

54 Marsh and Stoker (1995:17); see also Beckman (2005:9).

Instead, I devote a chapter to each of the five aforementioned dimensions of the theoretical framework. By structuring the written account around the *world view*, the *core values*, the *organizational ideals*, the *means of mobilization* and the *target group* of regionalism, I hope to foreground the aim of the work, namely, to gain a better understanding of the value-basis of regionalism in contemporary, less favoured and politically defined regions.

Thus embedded, the comparative approach to studying ideologies has assisted me in my attempts to identify commonalities as well as divergences between the cases. Many of the aforementioned models of ideology analysis have in fact been worked out in order to facilitate comparisons between different ideologies. Here, however, I suggest that the model works just as well when one aims to compare cognate ideologies across space. Before I present the empirical findings, the two case regions will be introduced in more detail. These presentations revolve around central economic, political and cultural aspects in the regions and will, accordingly, provide the study with a much needed general context. The comparability of the cases will be considered and the advantages of a comparative approach will be discussed in more detail.

3. The Cases: Comparing Norrbotten and Mecklenburg-Western Pomerania

Much of the essence of comparative social sciences has been understood to lie in the selection of cases.[1] One prominent idea has been that comparisons are to be made either between cases that are similar or between cases that are different.[2] The ideal in a *most similar* systems design is that the units analysed (whether countries, regions, institutions or actors) are equal in as many respects as possible. In an ideal *most different* case comparison, the cases studied should be different in as many respects as possible. The aim is then to find points and patterns of similar 'outcomes' between the cases that differ, and dissimilar 'outcomes' between the cases that are essentially similar. This procedure is thought to sort out the factors that seem to determine the outcome, and accordingly could be supposed to carry some sort of explanatory power.[3]

The idea of selecting cases according to their similarities and differences gives rise to a number of objections. A *first* objection is that every search for a single independent variable that determines the outcome of a case will be in vain. Political life involves the interaction of so many forces that, as Peters puts it, any 'artificial isolation of the casual factor would almost certainly be misleading as well as less interesting'.[4] A *second* objection would be that a selection of cases according to their similarities and differences runs the risk of being biased. To establish whether two cases are more similar than different requires a substantial familiarity with the cases – something that is commonly achieved only after a comprehensive study of them. The question of when a similarity is a similarity and when a difference is a difference is by no means simple to answer. What initially is perceived as a difference can be a 'functional equivalent', that is, a similarity in functions but a difference in the institutions and bodies executing

1 Landmann (2003:46).

2 Peters (1998:37); Marsh and Mackie (1995:178); see also Hague *et al.* (1992:23); Harle, Vilho (1989), 'Towards a Comparative Study of Peace Ideas: Goals, Approaches and Problems', *Journal of Peace Research*, 26(4), pp.341-351, p.344

3 Marsh and Mackie (1995:178).

4 Peters (1998:2,28-30).

them.[5] And what initially is perceived on the surface as a similarity can just as well be a difference in terms of functions and substance. Another argument in this respect would be that a comparison of two cases essentially is a 'most different systems' design. Thus, even if there are parallels between pairs of cases, society is so diverse that most of them also will prove to be substantially different.[6] While endorsing this idea, I regard the comparison between Norrbotten and MV as a comparison between cases that are essentially dissimilar, even if they share some certain characteristics. Yet, I endorse Sartori's argument that units need to share some properties in order to be at all comparable.[7] The case presentations in this chapter are accordingly thought to specify the properties that the cases share and that make a comparison feasible.

A comparison of discourses is – as indicated in the methodological discussion – not without problems. Discursive manifestations of ideology can indeed be compared in both time and in space, but only provided that they are considered in relation to their respective contexts.[8] Therefore, the presentations below are also meant to serve a contextualizing end. It should be observed that 'a context' is no fixed, given or objective category.[9] The 'making' of a context is, like any empirical observation or inference, 'coloured by the theoretical framework of the observer'.[10] The contextualization of a study must therefore be considered both subjective and positioned in relation to the subject matter. As my intention was never to write the history of the two regions in general terms but to explore the value-basis of regionalism and to understand recent processes of region building, the presentation of the cases will concentrate on factors that are relevant to these issues. The contextualization provided here will therefore be less detailed and less 'thick' than in a classical descriptive-inductive study.

---

5 Larsson, Reidar (1997a), *Jämförande politik*, Lund: Studentlitteratur, p.14; see also Ragin, Charles C. (1994), *Constructing Social Research*, California: Pine Forge Press, p.85.

6 Peters (1998:66).

7 Sartori, Giovanni (1991), 'Comparing and Miscomparing', *Journal of Theoretical Politics*, 3(3): pp.243-257, p.245.

8 Ragin and Hein (1993:270); see also Landmann (2003:43-44); van Dijk (1997a:3).

9 van Dijk (1997b:15f).

10 Marsh and Stoker (1995:15).

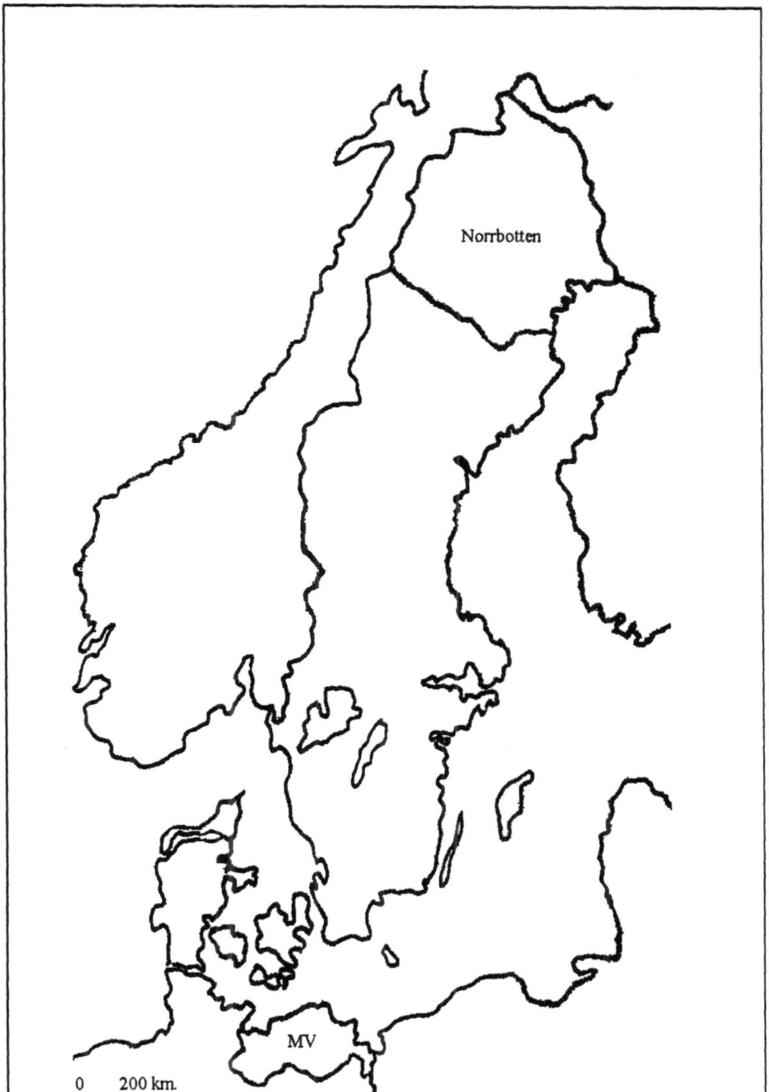

Map 1. The Baltic Sea Area

NORRBOTTEN, SWEDEN

Norrbotten is one of 21 counties in Sweden and was constituted in its contemporary form in 1810. The county covers about a quarter of Sweden's total land area, which makes it Sweden's largest county by area. However, the county hosts only about a quarter of a million inhabitants and proves in consequence to be an extremely sparsely populated region.[11] There are indeed considerable differences between the municipalities in the region when it comes to population density and migration flows. The population density is highest in the coastal areas in the south-east, in the north-western mining districts and in the lower river valleys.[12] Besides, migration flows to and from Norrbotten altered considerably during the 20th century. The population numbers almost doubled during the first half of the 20th century, not only because of migration but also because of an excess of births over deaths.[13] Thereafter, one has observed periods of increase as well as decrease of population. Since the late 1960s, however, the Norrbottnic population has decreased by about 6,600 inhabitants.[14]

A Less Favoured Region

The negative migration flows from Norrbotten are clearly linked to the limited regional labour market. The Norrbottnic economy has by tradition relied on primary industries such as forestry and timber, mining and steel. These branches have been considered important for the industrialization and modernization of Sweden as a whole, and are said to have moulded Norrbottnic society in many ways.[15] These primary industries have undergone extensive structural change during recent decades, with rising unemployment as one consequence. Between the mid-1980s and 2001, the number of employees decreased by 23 000.[16]

---

11 For a discussion of definitions of sparsely populated areas, see Eckerberg, Katarina and Wide, Jessika (2000), *Actors, Institutions and Attitudes to Rural Development: The Swedish National Report*, Umeå: Department of Political Science, Umeå University.

12 Erlandsson, Ulf (2005), 'Norrbotten', *Nationalencyklopedins internettjänst*; see also Pettersson, Örjan (1999), *Population Changes in Rural Areas in Northern Sweden 1985-1995*, Umeå: CERUM, Umeå University.

13 Erlandsson, Ulf (2005).

14 Lst Norrbotten (2004a), *Facts about Norrbotten 2004*, Luleå.

15 Sörlin, Sverker (1988), *Framtidslandet*, Stockholm: Carlssons, p.56.

16 Lst Norrbotten (2003a), *Fakta om Norrbotten*, Luleå, p.28.

Unemployment figures differ as between the coastal areas on the one hand and the inland municipalities and Finnish border regions on the other. Yet Norrbotten contains four of the ten municipalities in Sweden with the lowest employment rates.[17] There are also indications that public health in Norrbotten is worse than in other parts of Sweden.[18] On this matter, some argue that the extensive use of sickness benefits is an indication, not of public illness, but of the lack of other sources of livelihood.[19]

Cultural Context

The county of Norrbotten is, as already indicated, a political and administrative region established by the central state in 1810. To say that Norrbotten is a political region rather than an ethnically defined one does not imply that culture or ethnicity is without importance there. The borders of Norrbotten are drawn right across what have been perceived to be important cultural spaces. The border with Finland in the east crosses the so-called Torne Valley, an area in which the Finnish language by tradition has been very influential. Integration of the Torne Valley into Swedish society was an explicit governmental goal at the beginning of the 20th century, and the Finnish- and *Meänkieli-*[20] speaking minorities encountered repression as well as assimilationist policies on the part of the central government.[21] The question whether *Meänkieli* is a language or just a dialect of Finnish is still debated today. At the end of 1999, however, the Swedish parliament declared the *Meänkielis* to be an official national minority, and *Meänkieli* an officially recognized national minority language.[22] Since 1

17 Glesbygdsverket (2002)

18 Glesbygdsverket (2004), 'Ohälsodagar per invånare fördelade på skogslänen och samtliga län. 16-64 år. 1999-2003. Antal.', at *www.glesbygdsverket.se*

19 See NSD (2005), 'Saco-chef pekar ut norrbottningarna som arbetsskygga', 2005-04-22.

20 Literally translated, *Meänkieli* means 'our language'. *Meänkieli* is used synonymously with 'Torne Valley Finnish', that is, the Finnish spoken in the Torne Valley Area.

21 Lundén, Thomas and Zalamans, Dennis (2001), 'Local Co-Operation, Ethnic Diversity and State Territoriality – The Case of Haparanda and Tornio on the Sweden – Finland border', *GeoJournal*, 54(1), pp.33-42, p.35. For an extensive discussion about the swedification of TorneValley, see Elenius, Lars (2001), *Både finsk och svensk*, Umeå: Umeå Universitet.

22 Justitiedepartementet (2004), *Nationella minoriteter och minoritetsspråk - En sammanfattning av regeringens minoritetspolitik*, Ju 04.15; Stockholm.

April 2000, the right to use the Finnish, *Meänkieli* and Saami languages in communications with public administration has been protected by law in some Norrbottnic municipalities. The size of the *Meänkieli*-speaking minority is not clear, but figures from 20 000 to 50 000 individuals have been mentioned. Since ethnic categorization is due more to subjective than to objective criteria, drawing up a list of linguistic or ethnic minority groups remains problematic.[23]

The borders of Norrbotten have also been drawn right across an area that has traditionally been inhabited by the Saami. The Saami are best described as an indigenous ethnic and linguistic minority inhabiting what they call *Sápmi*, that is, northern Scandinavia and the Kola Peninsula. The number of Saami in Sweden has been estimated at around 20 000, but there is no official figure either of the total number of Saami or of the number of Saami in Norrbotten.[24] Just as in the case of the *Meänkielis*, there are methodological problems in estimating the Saami population. History has, moreover, been marked by ongoing conflicts between the ethnic Swedish majority and the Saami minority. Many Saamis have born witness to long-standing systematic oppression of the Saami minority, ranging from everyday resentments and racial prejudice to state-led attempts to eradicate Saami culture.[25] Over the years the Saami minority have been subjected to compulsory transfer of Saami villages, compulsion in religious matters and restriction of rights to reindeer pasture. According to Kvist, from the late 19th century to the early 1970s the Swedish state practised 'a system of institutionalized racism' towards the Saami minority. Kvist believes that the current policy shift towards ethnic tolerance is a result of the work of Saami organizations.[26] Nonetheless, the earlier discriminatory policy is said to have left 'a strong anti-Saami rights legacy among the non-Saami population of the North'.[27] Since 1993, there has been a Swedish Saami Parliament for whose members registered Saamis are entitled to vote. The Saami Parliament is not a self-governing body, but is rather to be described as a popularly elected public authority with strictly limited powers.

23 Elenius (2001:24,60).

24 Müller, D.K. and Pettersson, R. (2001), 'Access to Sami Tourism in Northern Sweden', *Scandinavian Journal of Hospitality and Tourism*, 1(1), pp.5-18, p.6.

25 Müller and Pettersson (2001:6).

26 Kvist, Roger (1994), 'The Racist Legacy in Modern Swedish Saami Policy', Umeå: Department of Saami Studies, Umeå University, p.203.

27 Kvist (1994:203).

Map 2. Norrbotten County

The County Council is the intermediate decision making level in Sweden. In modern times the Social Democratic Party (SAP) has always largely dominated the Norrbottnic County Council. During the time-period to which this particular study is limited, three general elections have been held (in 1994, 1998, and 2002). In all these elections, SAP won the largest share of the votes. After the 2002 elections, however, the SAP could form a majority only together with the Green Party and the Left Party. Since 1994 a regional party called the 'Health Care Party' has been represented in the county council.[28] The party has polled between 15 and 27 per cent of the votes in the elections, and is therefore a significant force in the county council.[29] Other parties represented in the county council are the Moderate Party,[30] the Liberal Party (*Folkpartiet*), the Centre Party[31] and the Christian Democrats.

The county councils are responsible for public health issues in the counties and for some other matters that are too costly for the municipalities to manage individually. Matters that prove to be of interest here are culture, tourism and those related to regional economic growth and development. Each year, the County Council publishes an Annual Action Plan[32] (AAP), in which the general visions and strategic aims of the Council's actions are set out. The AAPs establish the financial frames of these activities, but also identify preconditions and so-called 'important factors for regional success'.[33] All AAPs published between 1995 and 2001 have been taken into consideration in this study, and I have paid a particular attention to those parts of the documents devoted to

28 *Norrbottens Sjukvårdsparti* in the original. The party was founded in 1994, mainly as a protest against the political decision to close down the hospitals in Luleå and Boden and to replace them with a new one situated between the two cities.

29 In 2002, leading members in the Health Care Party and a local party called The Kiruna Party established a new regional party called Norrbottenpartiet (that is, the Norrbotten Party). The party ran for the National Parliament in the 2002 elections but unsuccessfully. It would have needed 12 per cent of the regional votes to enter the Swedish parliament, but managed to gather only 9 per cent of the regional votes. Since the election defeat, the party has been declared to be 'dormant'.

30 The party defines its ideology as a combination of 'a conservative outlook on society and liberal ideas'. See *www.moderat.se.*

31 The party changed its name from 'the Farmers Union' to 'the Centre Party' in 1958.

32 *Landstingsplan* in the original.

33 NLL (2005) 'Landstingsplan', at *www.nll.se/twosection.aspx?id=11134*, 2005-02-02.

regional development. Apart from AAPs, the county council produces reports, press releases, and opinion-forming statements within their fields of responsibility. The publications taken into consideration here have primarily concerned regional development and cultural and media policy.

The County Council's responsibility for regional growth and development is shared with the County Administrative Board – a central government agency with a twofold aim. The County Administrative Board is supposed to function as the representative of the Swedish Parliament and the central government at the regional level, but has also 'a parliamentary mandate to work for the county's best interests'.[34] Documents from the County Administrative Board have been taken into consideration primarily when they concern initiatives that are taken in cooperation with any other regional public institution. A County Governor, appointed by the central government, manages the County Administrative Board. Statements by the County Governor are here understood to aim partially at moulding regional public opinion. Accordingly, some of these statements have been taken into consideration, even though the County Governor is a public servant and not a popularly elected representative for the county.

In the mid -1990s the prevailing administrative and institutional structure in Sweden came increasingly to be seen as a problem. Apart from the County Council and the County Administrative Board, a large number of other regionalized public authorities are relevant in forming and implementing regional development policy. The situation was interpreted as a 'regional mess' and gave rise to a so-called 'regional experiment' that allowed a handful of counties in southern Sweden to establish different forms of regional bodies for self-government.[35] In a strict sense, the initial experiment concerned only this handful of counties, but in practice it came to influence the debate in many other Swedish counties.[36] Norrbotten – not part of the reform experiment – is no exception. One organization that has participated in the work for institutional reform in Norrbotten is the Regional Association of Local Authorities (RALA).[37] RALA has – partly in cooperation with the County Council – produced a number of reports about regional institutional reform. Some of these reports will also be referred to in this volume.

---

34 Sweden's County Administrations (2005), 'The County Administration and the Community', at *www.lst.se/english/society.htm*, 2005-06-20; Lst Norrbotten (2003b), *Perspectives on Norrbotten*, Luleå.

35 Jerneck and Gidlund (2001); Svenska Kommunförbundet, Landstingsförbundet (2002), *Att arbeta regionalt – ett nytt politiskt uppdrag*, Stockholm.

36 Statskontoret (1998), *Gamla län blir nya regioner?*, Stockholm, p.9.

37 *Kommunförbundet Norrbotten* in the original.

Apart from the actors mentioned above, we find some initiatives for cooperation in which questions related to regional development are discussed. Some of these initiatives go beyond the borders of Norrbotten County, but are still taken into consideration in the analysis. One such example is European Forum Northern Sweden, a partnership with the neighbouring counties aiming at positioning northern Sweden within the European Union. Reports from and debates within the Forum are referred to in some parts of this volume, as it is partially made up by elected representatives for the Norrbottnic municipalities and County Council.

MECKLENBURG–WESTERN POMERANIA, GERMANY

Mecklenburg–Western Pomerania (MV) is one of the 16 *Länder* of the Federal Republic of Germany, and a rather new political construct. Mecklenburg and Western Pomerania was part of the former German Democratic Republic (GDR) and was established as a political unit for the first time by the Soviet Military Administration in 1945. Only two years later, at the time of the dissolution of the Prussian state, the word 'Vorpommern' (Western Pomerania) was deleted from the name.[38] In 1952, owing to an administrative reform within the GDR, the *Land* was dissolved and subdivided into three districts. The establishment of Mecklenburg–Western Pomerania in its current form took place on 3 October 1990, the day of the unification of the former GDR and West Germany.[39]

A Less Favoured Region

The fact that MV once was part of the former GDR and nowadays is one of the so-called *Neue Bundesländer* is a matter of vital importance. Many analysts agree that the inhabitants of East Germany were displeased with the political

38 The reason for this was as follows. If one acknowledges that there is a 'Vorpommern', there must be a 'Hinterpommern', which implies that the two parts belong together. The idea of a united Pommern seemed reminiscent of Prussia, which in turn was considered to be synonymous with militarism and with the initial stage of the Nazi era. See Bahra, Hanne (1997), *Mecklenburg –Vorpommern*, Hamburg: Rowolth Verlag, p.11.

39 Kuhn, Heinrich-Christian (2000), 'Mecklenburg-Vorpommern – Gemeinsamkeiten trotz unterschiedlicher Geschichte', in Wehling, Hans-Georg (ed.), *Die Deutschen Länder – Geschichte, Politik, Wirtschaft*, Opladen: Leske + Budrich, p.143.

system of GDR, but that rather few of them were united around an alternative political project. Some wanted the GDR to become part of West Germany; others wanted a reformed GDR to persist.[40] Still, there were some expectations – *Wendeerwartungen* – that came to unite large parts of the population at the time of *die Wende*. Freedom of travel, democratic elections, freedom of opinion, and a more Western-style standard of living have been cited as widely shared expectations. In some limited respects, these expectations were at least formally fulfilled. The political rights demanded materialized to a large extent with the unification with West Germany, and the efforts made by the German Federal Government to improve the infrastructure and to renovate dilapidated buildings implied both job vacancies and a radical improvement of the street scene in many parts of eastern Germany. Due to direct subsidies from the federal government and to temporary regulations that aimed at stimulating investments in eastern Germany, this trade turned into something called a *Konjunkturlokomotiv*,[41] which managed to generate activities in other parts of the economy.[42]

Still, the structural transformation that took place in East Germany after *die Wende* had problematic consequences as well. With a population of scarcely 1.9 million inhabitants and an area of 23,170 km,[2] Mecklenburg and Western Pomerania were the most sparsely populated areas in the GDR.[43] Since *die Wende*, the population has decreased by about 160,000 inhabitants, primarily due to net migration losses. Given the higher west German population density, Mecklenburg–Western Pomerania is now a very sparsely populated area compared with the average of united Germany. One of the main reasons for the negative migration flows is the high unemployment in the years after the first unification phase. At the time of unification, the most important economic activities in Mecklenburg-Western Pomerania were shipbuilding and agriculture.[44] As these two activities branches belonged to two of the fastest-declining sectors in Europe at that time, MV entered the process of restructuring

40 Schröder, Richard (1993), *Deutschland schwierig Vaterland – Für eine neue politische Kultur*, Freiburg: Herder, p.35ff.

41 Figuratively, 'a force that contributes to an upward economic trend'.

42 Gurgsdies, Erik (1996) 'Vom Rand in die Mitte? Wirtschaftsentwicklung in Mecklenburg' in Friedrich-Ebert-Stiftung, Landesbüro Mecklenburg-Vorpommern (ed.) *Mecklenburg-Vorpommern Land am Rand – für immer?*, Schwerin.

43 Statistisches Landesamt at *www.statistik-mv.se*, 2005-06-02; Statistische Bundesamt, Deutschland at *www.destatis.de/basis/e/bevoe/bev_bsp_t2.htm*, 2005-06-02.

44 Pfeiffer, Ulrich and Dübel, Achim (1994) *Grundzüge einer wirtschaftlichen Entwicklungsstrategie für Mecklenburg-Vorpommern*, Schwerin: Friedrich-Ebert Stiftung.

from a highly disavantageous position.[45] In Western Pomerania alone, 80 per cent of those who worked in industry and agriculture lost their jobs after the unification.[46] Larger parts of eastern Germany are still struggling with severe economic problems. Approximately a quarter of the workforce in the former GDR suffers from 'open or hidden unemployment'.[47] The *Neue Länder* in general depend to this day on transfers from western Germany, and Mecklenburg–Western Pomerania is definitely no exception in this respect.[48]

Almost fifteen years after the initial stage of the German unification, the alleged 'mental wall' between East and West Germans has proved to be a topic of continuous debate. The question whether there is a common German national identity has been of constant interest since the mid-1990s.[49] Some have argued that East Germans – *Ossis* in popular speech – still feel like *Bürger zweiter Klasse*,[50] that they experience 'relative deprivation' and that they have 'lost self-respect' after the German unification.[51] One important aspect of this matter is that the transformation of East Germany involved a large-scale transfer of civil servants, bank clerks and university staff from West to East Germany.[52] Some have understood this process as a West German 'colonization' of East Germany.[53] Yet others have questioned the extent to which the personnel transfer of West German experts really pushed East Germans out of the political arenas.[54]

---

45 Gurgsdies (1996).

46 Werz, Nikolaus and Schmidt, Jochen (eds.) (1998), *Mecklenburg-Vorpommern im Wandel – Bilanz und Ausblick*, München: Olzog Verlag, p.14.

47 Hefeker, Carsten and Wunner, Norbert (2003), 'Promises Made, Promises Broken: A Political Economic Perspective on German Unification', *German Politics* 12(1), pp.109-134, p.110.

48 Hefeker and Wunner (2003:110).

49 Ekman, Joakim (2001), *National Identity in Divided and Unified Germany*, Örebro: Örebro University; Lutz, Felix Philipp (2002), 'Historical Consciousness and the Changing of German Political Culture', *German Politics* 11(3), 2002, pp.19-34.

50 'Second-class citizens'.

51 Schröder, Richard (2000), 'Zehn Jahre deutsche Einigung', *Das Parlament*, 2000-09-29; Kim, Mi-Kyung and Robertson, John D. (2002), 'Analysing German Unification: State, Nation and the Quest for Political Community', *German Politics*, 11(3), pp.3-18.

52 Maier, Jürgen (2003), 'Political Culture in East and West Germany', *Bamberger Beiträge zur Politikwissenschaft,* Nr. II-14, Universität Bamberg, p.4.

53 For a discussion on this, see Maier (2003:4).

54 Wollmann, Hellmut (2002), 'Local Government and Politics in East Germany', *German Politics*, 11(3), pp.153-178, p.155.

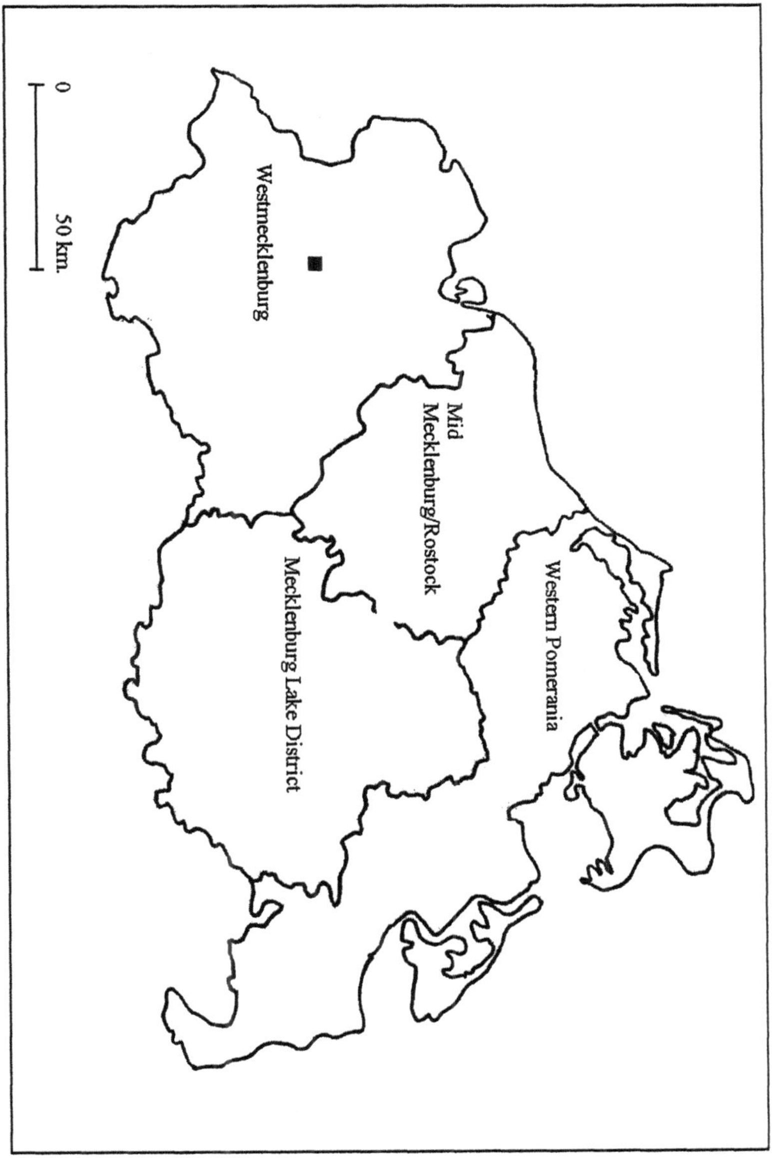

Map 3. Mecklenburg-Western Pomerania.

Like Norrbotten, MV is a politically defined region. The tensions between the ethnic Swedish majority and the indigenous minorities lack an equivalent in Mecklenburg–Western Pomerania. Still, the features mentioned above – *die Wende*, the unification, the establishment of new political units and economic struggle – have all been put forward as explanations for one of the most dramatic events in the contemporary history of Mecklenburg–Western Pomerania: the pogrom against asylum-seeking refugees, immigrants and foreign guest workers in Rostock–Lichtenhagen in August 1992.[55] Right-wing violence against asylum seekers had occurred in MV before the events in Lichtenhagen, but never to such a great extent.[56] Hence, during the entire summer of 1992, people from suburbian Lichtenhagen had complained and protested against the asylum house *Sonnenblumenhaus*. The protests concerned both the neglected maintenance of the building and its Vietnamese inmates. When these protests escalated in late August into a malignant demonstration attended primarily by local and visiting neo-Nazis, many in the neighbourhood came out to support it. The demonstration went on for several days, and when the *Sonnenblumenhaus* – inhabited by roughly 100 Vietnamese asylum seekers – was set on fire by the mob, a large number of spectators were reported to have applauded and cheered on the attack.[57]

The events in Lichtenhagen contributed to strengthening the idea of East German inhabitants as potentially right-wing. Ever since *die Wende* a propensity for right-wing extremism has been part of the image of the eastern parts of Germany in general and of Mecklenburg-Western Pomerania in particular.[58] Some have suggested that the alleged xenophobia in the eastern parts of Germany can be explained by the restricted contact with foreigners during the time of the GDR.[59] The share of foreign immigrants in Mecklenburg–Western

---

55 Betz, Hans-Georg (n. d.), 'Perspectives on Right-Wing Extremism', Canadian Centre for German and European Studies York University, Toronto at *www.aicgs.org/at-issue/ai-belz.shtml* 2004-03-28; Schröder (1993:29-30); Werz, Nikolaus (2000), 'Politische Reaktionen auf den Rechtsextremismus in Mecklenburg-Vorpommern', in Werz, N. and Hennecke, H.J (eds.) *Parteien und Politik in Mecklenburg-Vorpommern*, München: Olzog Verlag, p.170ff.

56 Werz (2000:170ff).

57 Dien Hong (2005), *Vereinsgeschichte*, Rostock; Living with Conflict (2003) 'Von Menschen, Ansichten und Gesetzen' at *www.living-with-conflict.org/v03/cms.php?ID=602&language=DE*, 2003-11-21

58 Werz and Schmidt (1998:13).

59 Schröder (1993:34).

Pomerania is indeed among the lowest in contemporary Germany. Still, the events in Lichtenhagen gave rise to a popular protest against racism. Right-wing extremist parties such as the *Nationaldemokratische Partei Deutschlands* (NPD) or *Deutsche Volksunion* (DVU) have, contrary to what has been the case in other *Neuen Länder*, had no success in elections to the MV *Landtag*.

Relevant Political Actors

After the first post-unification elections in October 1990, the Christian Democratic Union (CDU) and the Liberal Free Democratic Party (FDP) formed a coalition government that, apart from a reconstruction of the government in 1992, held the office until the 1994 elections.[60] As in Norrbotten, three general elections were held during the period of this particular study (in 1994, 1998, and 2002). After the 1994 elections, the CDU once again formed a coalition government. This time, the coalition partner was not the FDP but the Social Democratic Party (SPD). After the 1998 elections, Mecklenburg–Western Pomerania turned out to be the first *Land* in which the SPD joined a coalition government with the Party of Democratic Socialism (PDS). The SPD decision to accept the PDS as a coalition partner gave rise to a number of protests, as the PDS is the successor to the East German Communist Party (*Sozialistische Einheitspartei Deutschlands*).[61] In the 2002 elections, both parties won a renewed vote of confidence, and they were able once again to form a coalition government.

The parties that have been taken into consideration in the analysis are first and foremost those that have been represented in the *Landtag* during the latter half of the 1990s (that is, PDS, SPD, CDU). There are, however, also regional organizations for both the Green Party and the FDP in the *Land*, and in a minority of cases references are made also to these parties' publications. As in the case of Norrbotten, party programmes, election manifestos, press releases, and campaign material have been included in the analysis. As the political administration in Mecklenburg–Western Pomerania is more extensive than in Norrbotten, access to documentary sources has to some extent been better here. The aforementioned written records have thus been complemented with transcripts of speeches, petitions, motions, and debates from the *Landtag*.

---

60 Werz and Hennecke (2000:5).

61 O'Loughlin, J., Witmer, F. and Ledwith, V. (2002), 'Location and Political Choice in Post-Unification Berlin', *Eurasian Geography and Economics* 43(5), pp.349-382.

In analyses of governmental records, special attention has been paid to the ministries in charge of regional developmental issues, cultural policy and questions related to European and external affairs. The ministries' respective fields of responsibility have altered since the mid-1990s, but in principle it is the State Chancellery, the Ministry for Cultural Affairs[62] and the Ministry for Regional Development[63] that have been of greatest interest here. Some of the responsibility for spatial planning and regional development issues is in MV delegated to a sub-regional level, that is, to the four Regional Planning Associations (RPAs).[64] The RPAs have the districts and larger municipalities of the sub-region as their members, and their primary task is to work on spatial planning.[65] The RPAs are also formulating regional development strategies and *Leitbilder*[66] for sustainable development in their respective sub-regions. In these strategies, the alleged strengths and weaknesses of the sub-regions are analysed, visions of future development are drafted, and strategic policy recommendations are framed. Since these documents are very much linked to the political discussion at the level of the *Land*, they have been of vital interest to the analysis here.

COMPARABILITY OF CASES

I agree with the argument that, even if there are parallels between pairs of cases, society is so diverse that most such pairs will also prove to be substantially different.[67] And I do agree with the idea that units can be 'comparable with respect to ... the properties they share, and non-comparable with respect to the properties they do not share'.[68] In this study, both cases share the properties of being politically defined, less favoured regions. Both regions suffer from

---

62 *Ministerium für Bildung, Wissenschaft und Kultur in the original.* MfBWK for short.

63 *Ministerium für Arbeit, Bau und Landesentwicklung in the original.* MfABL for short

64 The institutional set-up in Mecklenburg–Western Pomerania has been under discussion in a later stage of this work, but during the time-period to which this study is confined there have been four Regional Planning Associations (*Regionale Planungsverbände*). See Map 3.

65 MfABL MV (2000e), *Spatial Planning in Mecklenburg-Vorpommern*, Schwerin, p.3.

66 'Lodestar'.

67 Peters (1998:66).

68 Sartori (1991: 245).

unemployment higher than the national average, and negative migration flows prove to cause problems in both regions. Concerning access to authoritative decision making, the two regions still prove to be essentially different. Norrbotten is part of Sweden, a unitary state in which the regional political level is less influential than both the local and the central levels.[69] Mecklenburg–Western Pomerania is part of Germany, a federal state in which the regional level, to the contrary, is among the strongest in the European Union. Thus, Germany has been understood as a 'prototype of cooperative federalism', originally designed to hinder the 're-emergence of a strong central state'.[70] This implies that the legal capacities of the regions compared are essentially dissimilar. The *Länder* have their own constitutions, elected parliaments, and governments, and the right to legislate on a variety of matters. The County Councils are, by contrast, responsible primarily for public health and medical services at the regional level, and for those matters that are of common interest to the municipalities but too costly for them to manage individually.

More important, however, is the fact that both regions host policy-making institutions that produce strategies, programmes and plans that focus on the development of the region. Thus, a single region may be one of many instances in a population of regions, but should never be understood as an actor in itself. The actors in this study are all linked to the policy-making institutions in the regions, to which the reader has been introduced. Still, it is the *narratives* of these actors, rather than their *functions* that are to be analysed in later parts of this volume. The ambition here is to compare and draw conclusions about the value-base of regionalism and region building, not to list the differences between institutional orders or legal capacities in two political entities. Therefore, it is the very content of the documents produed by the actors that is in focus here.

Reasons for Comparison?

The initial discussion about similar and different cases indicated in a sense that most comparisons aim to find one or more factors that can explain the outcome of a certain case. This is, however a misconception, as comparisons can serve many other analytical purposes. A central argument for a comparative approach is its capacity to generate, advance, or complement theory.[71] A comparison of a

69 Olsson, Jan and Åström, Joachim (2003), 'Why Regionalism in Sweden', *Regional and Federal Studies*, 13(3), pp. 66-89, p.69.

70 Börzel (2002:45).

71 Ragin (1987:34); Landmann (2003:4-5); Denk (2002:19).

small number of cases can advance theoretical thinking, not by allowing generalizing explanations but by enabling conceptual development and elaboration.[72] The in-depth knowledge a case study comparison can offer provides us, as Ragin and Hein argue, with a 'rich raw material for advancing theoretical ideas' and with the opportunity to illustrate broader perspectives and theories.[73]

Another argument for a comparative approach emanates from its capacity to serve contextualizing ends.[74] In analysing the same matter in different settings, the researcher is more or less forced to reflect upon the contexts in which the matter in question works. Besides, an analysis of 'several instances of the same thing' can make differences and similarities between cases more visible.[75] A comparison can thereby help us avoid simplification and lessens the risk of both false particularization and false universalism.[76] This holds in particular for comparisons between a small number of cases since, according to Sartori, they 'sacrifice generality to depth and thickness of understanding', whereas a comparison of a larger number of cases 'sacrifices understanding-in-context'.[77]

In this work, it is assumed that a comparison of *policy discourses* may help to refine our understanding of what the central concepts in the regional developmental debate are.[78] Thus, the most probable outcome is that different political actors with diverse social and institutional positions give voice to different beliefs, norms and values. If we instead discover that actors in dissimilar places with no explicit relation to each other share a similar discourse, we need to discuss the reason for these similarities. In this matter, Peters has argued that the logic of comparative analysis is to search for similarities between cases that are substantially different.[79] Ragin also points out that the 'looking for similarities in unexpected places' may help social scientists to 'develop new insights that advance theoretical thinking'.[80] Here, the eventual identification of similarities in Norrbotten and Mecklenburg–Western Pomerania is meant to bring us to a better understanding of regionalism and region building.

---

72 Ragin (1994:84); Ragin and Hein (1993:266)
73 Ragin and Hein (1993:266,270); Ragin (1994:84). See also Harle (1989:342).
74 Landmann (2003:4).
75 Ragin (1994:86).
76 Marsh and Mackie (1995:174); Landmann (2003:4-5).
77 Sartori (1991:253).
78 Harle (1989:342).
79 Peters (1998:66).
80 Ragin (1994:85).

4. The World-view of Regionalism

As already indicated, different ideologies tend to host different attitudes towards the surrounding world. The same seems to apply to different kinds of regionalisms. Some regionalist movements of the 1960s and 70s have been attributed with a reluctant or even distrustful attitude towards internationalization and other external influences, whereas the archetypical new regionalism has been characterized rather by cosmopolitanism and readiness to face global competition.[1]

The regions analysed here do not belong to the group of regions in which ethnic separatism or claims for autonomy have dominated the political agenda. Neither do they belong to the successful 'new' growth-regions in the centre of the European Union. This raises the question of the kind of world-view political representatives of these two regions express. How do they conceive of social and political reality when making policy decisions and judgements? Are regions thought to be oppressed by a dominant state, or is the state thought to be declining? Is the region thought to be threatened by modernization, globalization and European integration, or are these processes thought to be advantageous for the region? These questions are central, as an enhanced understanding of how regional representatives comprehend and explain the world may help us put other aspects of regionalism into perspective.[2] The core values and the organizational ideals adopted, the symbolism used and the target group aimed at must all be understood in relation to the particular perception of social and political reality they proceed from.

Norrbotten

Globalisation

Globalization, European integration and the relation to the state are recurring issues in Norrbottnic policy discourse. Globalization is usually referred to as a

1 Harvie (1994:2).
2 Rejai (1995:4).

'remarkable change', as an 'ever faster process', and as 'the main driving force of our times'.[3] This is the case also with *The Guide*,[4] a publication that exerted a great influence on the regional development debate in Norrbotten during the 1990s. *The Guide* originates from a regional developmental project that was initiated and run by the Norrbottnic County Council (NLL), the County Administrative Board and the municipalities in the county during the first half of the 1990s. The parties mentioned took part in a joint analysis of the situation in the county and published the result in a booklet titled *At the Crossroads*.[5] The publication was sent out to all households in the county together with a questionnaire in which the inhabitants were asked to give their opinions on the future development of the region. After responses were obtained from the inhabitants, *At the Crossroads* was reworked, renamed *The Guide,* and adopted by the municipalities and a number of other public actors in the region.[6]

The message in *The Guide* is that escalating internationalization provides the region with new opportunities, but also with an accentuated competition. Globalization is understood to bring the region and 'the world' closer to each other. 'Step by step', it is argued, 'the region becomes part of something bigger', since the surrounding world gets 'more interested in us, and we take more of an interest in others'.[7]

Since *The Guide* was published, several actors have produced documents that refer to globalization, the alleged dismantling of the state and increased competition between regions.[8] Internationalization in quite general terms is commonly referred to as 'the most significant change and challenge of our times', and held to to an ever-escalating transformation process.[9] A number of

---

3 SAP Norrbotten (1997) *Tillväxt i Norrbotten – av egen kraft, Regionalt partiprogram*, Arvidsjaur, p.9; NLL (1998a), 'Angående *Handlingsprogram för det internationella arbetet i Norrbottens län'*, Dnr LK A1231-98, Luleå.

4 *Vägvisaren* in the original. Partnerskap för Norrbotten (1995), *Vägvisaren – Så här ska vi göra för att utveckla Norrbotten*, Luleå.

5 *Vid Vägskälet* in the original.

6 The document was agreed to by various actors at various points in time during spring and early summer 1995. Regionutredningen Norrbotten (1999a), *Regionalt Samarbete i Norrbotten – en återblick.* Delrapport 2, Luleå, p.4.

7 Partnerskap för Norrbotten (1995:4,18).

8 Vänsterpartiet Norrbotten (1998), *Länsprogram 1998-2002*, Luleå, p.4; Folkpartiet Liberalerna Norrbotten (1998), *Ny kurs för Norrbotten. Handlingsprogram 1998-2002*, Luleå p.7.

9 NLL (1994b), *Landstingsplan 1995*, Luleå, p.4,6; NLL (1995), *Landstingsplan 1996*, Luleå, p.6; NLL (1998c) *Landstingsplan 1999*, Luleå, p.6; NLL (1999:8); SAP Norrbotten (1997:2); SAP Norrbotten (2001) *Regionalt partiprogram*, Norr-

publications distributed by the County Council during the 1990s refer to the idea that Norrbotten and other regions in Europe are gradually becoming part of 'something bigger'. In these publications, we learn that the present time is characterized by a number of trends that undermine state authority and provide the region with new opportunities. Structural changes, migration flows and general 'centrifugal forces' are thought to be adding up to a gradual dismantling of the nation state and a renaissance of the regions.[10] In a report by the Regional Association of Local Authorities (RALA) and the County Council, the nation state is explicitly identified as the prime loser in 'a world dominated by global network societies'.[11] In other contexts, Sweden's membership of the EU is believed to have increased the importance of political and economic trans-border cooperation between regions.[12] Globalization or Europeanisation, however, are not the only factors cited as important for the region. In some cases, it is the fall of the Soviet Union, the disarmament at the Kola peninsula or closer contacts in the Barents region that are emphasized in accounts of the environment of the region.[13]

Regions More Important

We do hear voices arguing that Norrbotten is exploited, oppressed or neglected by 'Stockholm' or 'the rest of Sweden'.[14] Still, globalization is quite generally

---

botten, p.5; Wikman, Leif (2003), 'Norrbotten – en framtida tillväxtregion eller ett län under avveckling?', *NSD*, 2003-08-09.

10 NLL (1997a), *Vi arbetar för välfärd och utveckling*, Luleå, p.3; NLL (1996b), *Mediefakta 1996. Historik, Statistik och Utvecklingstendenser*, Luleå, p.7; NLL (1994a) *Kulturen för Norrbotten framåt - Kulturpolitiskt program*, Luleå, p.3, 29.

11 Regionutredningen Norrbotten (1999b), *Regionaliseringens drivkrafter - Delrapport 4*, Luleå, p.3.

12 NLL (1998c:6); NLL (1999) *Landstingsplan 2000*, Luleå, p.8.

13 Folkpartiet Liberalerna Norrbotten (2002a), *Ett län för individer! Valmanifest*, Norrbotten, p.6; Vänsterpartiet Norrbotten (1998:2); Holmqvist, Per (1998a), 'Norrbottens Sjukvårdsparti – en Regional Kraft', *Hjärtslag*, Specialnummer, p.3; Ohlsson, Staffan (1998a), 'Regionfullmäktige – Nödvändigt för demokratin', *Hjärtslag*, Nr 4, p.16; Zakrisson, Kristina (2003), 'Norrbotten behöver inga allmosor', *NSD*, 2003-05-28; Norrbottengruppen (1999) *Regionalt Tillväxtavtal för Norrbottens län 2000-2002*, Luleå, p.38.

14 Backgård, Kenneth (1998a), 'Mot Nya Djärva Mål', *Hjärtslag*, Nr 4, p.2; Norrbottens Sjukvårdsparti (1998), *Regionalpolitik*, Norrbotten, p.2; Holmqvist (1998a:3); Centerpartiet Norrbotten och Västerbotten (2001), 'Den Norra Vreden' *Gemensamt*

thought to have weakened the capacity of the state, which, in turn, is said to have resulted in an entirely new conceptualization of regions and regional policy.[15] In part, this development is perceived as a problem, primarily because regions 'cannot expect to get the same support from the state as before'.[16] Due to diminishing support from the state, it is argued, the region must become competitive and capable of providing for its own survival.[17]

The alleged impairment of the state's ability to care for the regions, however, is not understood as something entirely bad. Globalization or internationalization is thus represented as a potential for development rather than as a threat to the region.[18] The decline of the central state is thought to lead to a more prominent position for regions in general terms. In a report from the County Council and the RALA, references are made to a 'comprehensive body of research, showing that the regions are growing in strength when the nation state is losing ground'. According to this research, it is argued, globalization tends to walk hand in hand with a regional 'rebirth'. Globalization and state decline are then said to turn regions into the 'new carriers of Europe's distinctive character – multiplicity and variation'.[19] Here, as well as in other publications, European integration is thought to have enabled a 'Europe of the Regions', or, at least, a Europe in which the regional level has gained more decision-making power and more responsibility for its own development.[20]

Several actors, furthermore, stress that the state lacks not only the financial resources but also the knowledge and the actual capacity to generate long-term economic development at the regional level. We find representatives of socialist as well as of non-socialist parties arguing that 'traditional state subsidies' or the 'old subsidy mentality' are increasingly outmoded. Development, several actors argue, must start out from below, 'arise in cooperation between people' and grow by means of initiatives taken at the regional level.[21]

---

uttalande, Piteå, p.1; Abrahamsson, Olov (2002), 'En fråga om solidaritet', *NSD*, 2002-03-14.

15 NLL (1997c), *Landstingsplan 1998*, Luleå, p.6; NLL (1996b:7).

16 Partnerskap för Norrbotten (1995:8); Vänsterpartiet Norrbotten (1998:3).

17 Partnerskap för Norrbotten (1995:28).

18 Rosengren, Björn (1998), 'Förord' in Norrbottengruppen (eds.), *Internationell Strategi för Norrbottens län*, Luleå.

19 Regionutredningen Norrbotten (1999b:3).

20 NLL (1997c:6); NLL (1996b); NLL (1996b:7); Regionutredningen Norrbotten (1999b); Norrbottens Sjukvårdsparti (1998:3).

21 Folkpartiet Liberalerna Norrbotten (2002a:5); Centerpartiet Norrbotten (2002b), *Valmanifest*, Luleå; Centerpartiet Norrbotten och Västerbotten (2001:1); Öberg, Maria (2002), 'Glesbygdskommunerna behöver samverkan', *NSD*, 2002-06-26;

The era of governmental subsidies is understood to be definitely over, and the region is said to be taking over more and more of the responsibility for regional development.[22] We learn that regional growth and development 'to a greater extent than before will depend on ourselves', that the region needs to be better at 'maintaining its interests' and at 'taking care of the advantages that internationalization will bring'.[23] In a report from the County Council, published in the mid 1990s, these circumstances are said to increase the need for 'regional emancipation'.[24] In another context, this development is said to walk hand in hand with a growing regional consciousness, which is referred to as a result of a 'popular dissatisfaction with the achievements of the central state'.[25] Thus, the alleged failures of the central state are claimed to have increased regional citizens' willingness to 'take responsibility for their own development'.[26]

Regional Assets

One argument that seems to unite great sections of the regional policy-making elite in Norrbotten is that increasing competition amongst regions requires 'new political thinking'.[27] In a number of publications, it is declared that we (the Norrbothnians) 'cannot follow the same old route as before', that (we) 'need to renew ourselves', 'do a bit of rethinking' or follow new lines of thought.[28]

Sundström, Anders (2004), 'Norrbotten har fått en verktygslåda', *NSD*, 2004-04-13; Henriksson, Jan (1998), *Regional förlagsverksamhet och stöd till bok- och medieproduktion*, Luleå: NLL, p. 2; NLL (1994a:29).

22 See NLL (1998c:6); NLL (1997c); NLL (1999:5); Regionutredningen Norrbotten (2000b) *Slutrapport*, Luleå, pp.2f; NLL (2000a) *Kultur ger Norrbotten kraft*, Luleå, p. 4.

23 Partnerskap för Norrbotten (1995:4,8); SAP Norrbotten (1997:2f); Norrbottens Sjukvårdsparti (1998:3).

24 NLL (1996b:7).

25 Regionutredningen i Norrbottens län (1999d:6), (2000:3).

26 Regionutredningen i Norrbottens län (1999d:6).

27 Kommunförbundet Norrbotten (2000), *Tiderna förändrar oss*, Luleå, p.5; Oskarsson, Eva (1998), 'När partipiskan viner', *Hjärtslag*, Nr 4, p.9; Norrbottengruppen (1998) *Internationell Strategi för Norrbottens län*, Luleå, p.6; NLL (1996c), *Mediepolitiskt program*, Luleå, p.5; Regionutredningen Norrbotten (2000b:3).

28 Partnerskap för Norrbotten (1995:4); NLL (1997a:3); Regionutredningen Norrbotten (1999d:5); Gröning-Degerlund, Lotta (2001a), 'Landstingspolitikerna ett sammansvetsat gäng', *NSD*, 2001-09-27; Gröning-Degerlund, Lotta (2003b), 'Svar direkt', *NSD*, 2003-07-07.

Another recurrent argument is that the regions own resources, 'endogenous potentials' or 'soft development factors' that are of central importance for regional development. Factors such as 'human capital', 'knowledge', 'networks and milieus' and 'civic or enterprising spirits' are thus supposed to be decisive for the future development of the region.[29] Most parties in Norrbotten appear to stress that 'a motivated population', 'will, creativity and self-confidence', 'an attractive local and regional milieu' or the capacity to 'mobilize the own resources' are significant factors for regional economic development.[30] Many of these so-called 'soft factors' are indeed quite hard to define. Still, it is widely assumed that they constitute potential competitive advantages, and that they are as important for the development as ever the access to fixed assets or other means of production.[31]

The conviction that 'soft' factors such as those referred to will decide the future of the region is bound up with a conception of the world as a 'knowledge or information society'. In some contexts it is argued that such a society requires an increased capability to cope with changes.[32] In some of the action plans from the County Council, we learn that the 'evolving information society affects regional, economic and social conditions' and that 'knowledge has never grown at a faster rate than it does today'. 'Openness to new impulses', 'flexibility' or the ability to 'respond to the ever faster changes in the world' are in other cases emphasized as decisive for the future development of the region. Since these capacities and the 'ability to cooperate' are said to unite 'most regions of growth', they are understood as 'preconditions if Norrbotten is to be an attractive and competitive region'.[33]

Furthermore, it appears to be widely assumed that Norrbottnic society, in order cope with the ever-increasing global competition, needs to work on its overall *attitude* to development. Social democrats as well as liberals and conservatives in the county announce that they would like to see a more positive

29 Centerpartiet Norrbotten (1998), *Tillsammans lyfter vi Norrbotten, Länsprogram 1998-2002*, Luleå, p.10; Norrbottengruppen (1999:17,23).

30 Centerpartiet Norrbotten (1998:6,10); Folkpartiet Liberalerna Norrbotten (1998:6); Folkpartiet Liberalerna Norrbotten (2002a:5); SAP Norrbotten (1997:3); Vänsterpartiet Norrbotten (1998:4); NLL (1994a:3f,17,29); NLL (1997c:6); Regionutredningen Norrbotten (1999b:3); NLL (2000a:4); Norrbottengruppen (1999:17ff).

31 Norrbottengruppen (1999:23); SAP Norrbotten (1997).

32 SAP Norrbotten (2001:5); NLL (1997c:5).

33 NLL (1997c:5); NLL (1994b:6); NLL (1998c:7); NIR (2003b), 'Norrbottens internationella samarbete ska skapa resultat', *Pressinformation 2003-06-02*, Luleå.

attitude towards development in the region.[34] Voices from the County Council and from the County Administrative Board stress that a positive attitude, self-confidence and a willingness to look on the bright side are essential elements in a process of regional development.[35] It has also been said that factors such as the 'mentality', the culture and the history of the region are far more important for development than regional policies and strategies.[36] One recurrent argument in the documents referred to here is that 'the individual', 'the local community' or 'the Norrbothnian' have a great opportunity to decide their own future.[37] Hence, it is no longer external macroeconomic conditions or the state's willingness to support the region that are thought to determine the region's future destiny, but the capacities of the region itself. Regional actors need to learn to 'act rather than react', it is said, and thereby cease to be 'victims of circumstances'.[38]

Mecklenburg-Western Pomerania

Globalization

Given the end of the GDR and the subsequent German unification, it may not be surprising that in Mecklenburg-Western Pomerania (MV), too, we come across the argument that the world has become more globalized and competitive. Representatives of all established parties in the *Land* refer to globalization as a 'tremendous change', as a process that is going ever faster, and as 'the main driving force of our times'. The Party of Democratic Socialism (PDS), the Social Democratic Party (SDP) as well as the Christian Democratic Union (CDU) stress that 'our society experiences an era characterized by profound changes', that we are facing 'an increasing Europeanization and globalization' and that we experience 'a Europe in upheaval'.[39] In most cases, globalization is understood

---

34 SAP Norrbotten (1997:2); NSD (2002a), 'Det är jäkligt mycket gnäll', 2002-05-17; Hultin, Bo (2004), 'Samverkan och positivt tänkande ger tillväxt', *NSD*, 2004-09-10; Wikström, Åsa (2004), 'Det måste löna sig att tro på sin idé', *NSD* 2004-09-25.

35 Henriksson (1998:2); Eriksson, P.O (2002), 'Norrbottens möjligheter måste alltid framhållas', *NSD*, 2002-09-04.

36 Gröning-Degerlund (2001b), 'Kultur stämplas som hobby', *NSD*, 2001-09-28.

37 Henriksson (1998:2); NLL (1994a:29); NLL (1996b:7); NLL (1997a:3).

38 SAP Norrbotten (1997:2).

39 Blum, Andreas (1999), 'Neuen Herausforderungen Rechnung tragen, in PDS Fraktion Landtag MV (ed.), *Was macht Ihr da im Schloss*, Schwerin; SPD MV (2001) *Bildung schafft Zukunft - Leitantrag des SPD Landesverbandes MV*,

as an opportunity, but in others it is also interpreted as a great challenge.[40] In each case, however, the present era is understood as being 'new' in several respects and as such associated with new arguments and demands.[41]

When globalization and Europeanization are referred to, the fundamental challenge they pose is identified as increased competition rather than extended political responsibilities for the regional level. The CDU stresses in one of its publications that the *Land* needs to enter into competition with other regions in an offensive and 'aggressive' manner.[42] The SPD-led Ministry of Interior declares that the *Land* will need a modernized administration in order to 'keep up with other regions in the Baltic Sea' and with a Europe of the Regions.[43] At the sub-regional level, we learn that globalization and Europeanization make a distinctive image of the region an 'imperative necessity if the region is about to assert itself in the competition between regions'.[44] In one of the sub-regions, it is emphasized that the region has 'a sense of responsibility for its own development'.[45]

A New Geo-economic Position

Interesting in the policy discourse in MV, however, is that not only regions in general but this *Land* in particular are promoted as increasingly important. During the first years of the 21st century, the political debate in Mecklenburg-Western Pomerania was palpably influenced by the approaching enlargement of the European Union. With Poland's entry into the European Union, MV would no longer be situated on the outer border of the Union. Some voices certainly stress the allegedly peripheral position of Mecklenburg and, in particular,

Güstrow, p.1; CDU Landtagsfraktion MV (1996), *Identitätsgewinn im Aufbau Ost*, Schwerin, p.6.

40 StK MV (2002a), 'Chancen, Eigeninitiative, Zusammenhalt', *Rede, Minister Präsident Ringstorff auf FES-Tagung 2002-02-25*, Rostock; StK MV (1998) *Regierungserklärung zum Beginn der 3 Legislaturperiode*, Schwerin, p.7.

41 StK MV (1998:3-4); CDU Landtagsfraktion MV (2002), *Abwandern oder Anpacken? Mecklenburg-Vorpommern – ein Land mit Perspektive!* Schwerin, p.8-10.

42 CDU Landtagsfraktion MV (2002:10).

43 Innenministerium MV (2003), *Burgernähe statt Bürokratie*, Schwerin, p. 5.

44 RPV Mittleres Mecklenburg/Rostock (2002), *Regionale Agenda*, Rostock, p.6.

45 RPV Vorpommern (2001), *Leitbild für die Region Vorpommern*, Greifswald, p. 2.

Western Pomerania.[46] Yet, with the enlargement, the geo-economic position of MV came to be seen as much more advantageous than before.

The willingness to stress the allegedly new and central position of Mecklenburg-Western Pomerania may be bound up with the fact that this part of Germany in some contexts is talked about as a *Land am Rand*.[47] At all events, there seems to be a strong consensus around the importance of EU enlargement for the geo-economic and geo-political position of MV. The SPD-led State Chancellery stresses that, with enlargement, MV 'leaves its peripheral location behind and gains a central position in the Baltic Sea area'.[48] Also, the Social Democratic Party district declares that the enlargement of the EU 'brings new opportunities to the *Land*, since it will move from the fringe to the centre of the European Union'.[49] The PDS Minister of Regional Development maintains that enlargement will give MV the opprtunity to turn into a 'hub in the Baltic Sea area'.[50] In much the same manner the PDS party district stresses that with the enlargement of the Union, 'the *Land* has left its peripheral position behind'.[51] Similarly, the CDU argues that with enlargement MV is 'drawing into the centre of the second largest economy in the world'.[52] Within the next few years, it is argued, MV can turn out to be a 'modern' or 'successful' region in the 'the centre of an integrating Europe'.[53] Another geographical vision communicated by representatives from SPD, CDU and PDS alike, puts MV in the 'increasingly important' centre of the perceived triangle of Berlin, Szczecin and Hamburg.[54] In

46 RPV Vorpommern (1998), *Regionales Raumordnungsprogramm Vorpommern*, Greifswald, p.14-15.

47 i.e. 'Land at the edge'. See Werz and Schmidt (1998:12); Friedrich-Ebert Stiftung/Akademie Schwerin (1996) *Mecklenburg-Vorpommern, Land am Rand – für immer?*, Schwerin; Dettling, Warnfried (2001), *Perspektivenwechsel - Thesenpapier presented at Denkwerkstatt 2020*, Berlin.

48 StK MV (1998:26); StK MV (2003a), 'Auswärtige Angelegenheiten', at *www.mv-regierung.de/stk/eu-abt/konaus.htm*, 2003-05-12; StK MV (2002c), *Mecklenburg-Vorpommern – ein Land zum Leben*, Schwerin.

49 SPD MV (2002) *SPD – Die Kraft des Landes*, Schwerin, p.12.

50 Holter, H. (2000a), 'Die Osterweiterung: Chancen und Risiken für Mecklenburg-Vorpommern', *PDS Pressedienst* Nr 51/52/2000, p.21.

51 PDS MV (2002), *Wahlprogramm 2002*, Neubrandenburg, p.10.

52 CDU MV (2002), *Ein neuer Wille fürs Land – Wahlprogramm der CDU*, Schwerin, p.9.

53 CDU MV (2001), *Die Zukunftsoffensive der CDU: Politik gestalten – Mecklenburg-Vorpommern voran bringen*, Schwerin.

54 MfABL MV (2002c), 'Für Arbeit und Lebensperspektiven – Plädoyer für eine strategische Allianz', *Rede des Ministers Holter am treffen der Denkwerkstatt*,

yet another context the Minister for Cultural Affairs contends that 'our *Land* is located at the heart of Europe'.[55] Also at a sub-regional level, it is said that this part of Germany is developing into a part of the future Baltic Sea Area.[56]

Regional Assets

The world-view that stands out in regional policy discourse is based on notions about globalization, increased competition between regions, and the significance of the enlargement of the European Union. Besides, several actors appear to understand the modern era as 'the information age', as a post-industrial period or as an era characterized by a new economy.[57] As in Norrbotten, regional representatives are keen to argue that the region will depend more and more on its own resources, and that these resources cannot be fixed assets or primary products only. Here, just as in Norrbotten, so-called soft location factors are widely thought to be gaining in significance.[58]

Every established party in MV emphasizes the growing importance of the inner qualities, competencies and particularities of the region.[59] The SPD stresses that 'we are experiencing a transition from industrial society to a knowledge and information society' and that education, information and communication will gain constantly in importance.[60] The PDS Minister of Regional Development argues in a number of contexts that soft location factors are of crucial importance for regional economic success. As examples of 'endogenous potentials' or 'soft development factors', the Minister mentions 'attitudes and behaviours, attitudes to new technologies, inventiveness, entrepreneurial spirit and civic virtues such

---

Wolgast; Kaufmann, Christoph (2000), in RPV Meckleburgische Seenplatte (eds.) *REGIOnal,* No 3, Neubrandenburg; CDU MV (2001).

55 MfBWK MV (2002), 'Kulturentwicklung und kulturelle Leitlinien für Mecklenburg-Vorpommern', *Rede des Ministers Kauffold, 2002-06-05*, Schwerin.

56 RPV Vorpommern (2001:1).

57 SPD MV (2001:1); CDU Landtagsfraktion MV (2002:20).

58 CDU Landtagsfraktion MV (2002:8,12); RPV Vorpommern (1998:29,96).

59 See also Grüne MV (2002c), *Wirtschaftspolitisches Grundsatzpapier*, 2002-11-23, Rostock; Grüne MV (2002d), *Wirtschaftspolitische Leitlinien*, 2002-11-24, Rostock.

60 SPD MV (2001), *Bildung schafft Zukunft - Leitantrag des SPD Landesverbandes M-V*, Güstrow p.1; MfBWK MV (2003), *Integratives Leitbild – Bildung Wissenschaft und Kultur für die Entwicklung von Mecklenburg-Vorpommern*, 2003-12-03, Schwerin;

as openness and tolerance'.[61] It is also contended that cultural and social infrastructure, or features such as image and *Heimatgefühl*,[62] are qualities that would benefit regional economic growth.[63] Besides, things such as quality of life and 'non-material wealth' are also said to be gaining in importance for regional economic development.[64]

A publication from the CDU maintains that 'we need to gather everything in our *Land* that may help us to progress'. It also holds that development will be achievable only through 'tenacity, power, ideas and without any kind of false modesty'.[65] In other contexts, it has been held that the mere *will* of the regional population will be of decisive importance for a region's future prospects.[66] More common than this, however, is to ask for strategies capable of 'strengthening the strengths' of the region and of safeguarding and protecting its endogenous potentials. In fact, a policy that strengthens the strengths of the region and counteracts its weaknesses seems to one of the most widely agreed.[67]

A New Regionalist World-view?

The world-view communicated in the two regions studied is clearly influenced by 'new regionalist' thought. Thus, *first*, globalization is generally referred to as

---

61 MfABL MV (2000a), 'Nachhaltige Politik für die Zukunft Mecklenburg-Vorpommerns' *Begrüßung der Teilnehmer der Denkwerkstatt 2020*, 2000-04-28, Parchim.

62 'Sense of community'.

63 Holter, H. (2004b), 'Kräftige Impulse durch Initiative Baukultur im Land', *Grußwort zum Tag der Architektur 2004*, 2004-06-24, Rostock; Holter, H. (2003a), *Aktive Arbeitsmarktpolitik statt gefährlicher Flickschusterei, Rede im Landtag*, 2003-05-22, Schwerin.

64 Holter, H. (2004a), 'Die Städte stärken - Wandel erfordert neue Allianzen', 2004-03-18, Berlin.

65 CDU Landtagsfraktion MV (2002:10)

66 Rehberg, Eckhardt (2002), *Rede zu Wahlauftakt 23 März 2002*, Warnemünde.

67 See for instance RPV Mecklenburgische Seenplatte (2000a), *Regionales Entwicklungskonzept*, Neubrandenburg, p.1,18-48; Sinz, Manfred (2000), in RPV Meckleburgische Seenplatte (eds.), *REGIOnal*, No 2, Neubrandenburg, p.9; RPV Vorpommern (1998:29); Delheim, Judith (2002), ,'Regionales in Programdebatte und Praktischer politik der PDS', *Disput* 2, p.18; CDU Landtagsfraktion MV (2002:7-8); Grüne MV (2002d); Grüne MV(2002b), *Selbstbewußte ostdeutsche Perspektive*, 2002-02-21, Rostock; SPD MV (2002).

an opportunity and not as a threat.[68] Globalization is understood to bring new chances to most localities, given that the region (as a whole) adapts to the new situation and focuses on the factors that would strengthen the regional competitiveness. The largely positive representation of globalization may be a bit surprising. Thus, globalization is certainly a complex phenomenon that affects different localities and different groups in different ways. And in other contexts, political interpretations of these processes have been much more diverse.[69] *Second*, the state is said to be in decline, particularly by Norrbottnic political representatives. The state is no longer perceived as an homogenizing, centralizing force that threatens the distinctive features of the region. Neither is it referred to as an institution the region can address with claims for more subsidies. The decline of the central state is mostly referred to as something inevitable, and, although state decline is thought to lead to diminishing state support, it is also referred to as something predominantly positive. Regions are widely thought to gain in importance, but whereas policy-making in Norrbotten seem to proceed from the assumption that regions *in general* are becoming more important, policy-makers in MV tend to accentuate the increasing importance of this particular *Land*. In short, it is the region, not the state, the European Union or any other unit that stands out as the most important unit in society.[70] *Third*, the world is not only understood as increasingly globalized and competitive. It is also perceived as being more and more of a knowledge society. Given that, 'soft location factors' such as knowledge, creativity and flexibility are presumed to gain in significance in both Norrbotten and Mecklenburg-Western Pomerania.

Adoption of Phrases or Real Consensus?

Some may say that the world-view referred to here is completely appropriate or even the only possible stand for regional representatives to take today. And it can indeed be argued that the world *is* globalized, that the state *has* less capacity than before to support less favoured regions and that the we really *are* 'passing through a major transition … in … geopolitical development'.[71] This is also a

68 See Chapter 1.
69 Keating *et al.* (2003:13).
70 Larsson (1997b:20).
71 Gregory, D., Martin, R. and Smith, G. (1994), *Human Geography*, London: Macmillan Press, p.2.

conception of the world that recurs in academic work.[72] Still, there are voices – as 'academic' as the ones just mentioned – arguing that the state is in full vigour, although its powers are used in different ways from before.[73] The ambition here is yet not to decide on the 'best' approach to globalization or the most appropriate description of the capacity of the state. Thus, it is the political interpretation of an alleged reality that is of interest here – not 'reality' in itself. From that point of departure, we may ask why regions such as MV and Norrbotten host so few alternative interpretations of the current world-order. Why does the focus on competition and competitiveness meet with so little objection? Why is confidence in the prospects of becoming competitive so strong? Why are there so many discursive similarities across the regions and across party political and organizational borders?

One suggestion would be that what we have seen so far is nothing but 'eurojagon', that is, phrases that recur frequently in the EU bureaucracy and that are adopted by regional representatives simply in order to obtain finance. Both regions have indeed been targets of significant structural funding during the time-frame of this study, and thus a simple explanation of the similarities would be that the representatives in these regions have adjusted their official use of language to the demands of the EU. Interpretations of this kind have figured in the scholarly debate before. Thus, the formulation of Single Programme Documents and other documents related to EU measures can be regarded as processes that take place 'at the top level of the administrative hierarchy only'.[74] And in some cases, the drawing up of these documents can be considered having a 'defensive character of "filling in the answers" to the common European scheme'.[75]

Here, however, the suggestion is that such an explanation may be a bit superficial. Many of the documents referred to have been produced with no formal connection to EU Structural Funding. The audience for the vast majority of documents analysed here is thus not made up of the European Commission

72 Deas, Iain and Ward, Kevin G. (2000), 'From the 'New Localism' to the 'New Regionalism'? The Implications of Regional Development Agencies for City-Regional Relations', *Political Geography* 19, pp.273-292, p.276; Giddens, Anthony (1999), *Tredje Vägen – om förnyelsen av socialdemokratin*, Stockholm: Atlas förlag; Lash and Urry (2002:279-280).

73 Keating (1997:27); Keating *et al.* (2003:13).

74 Jacobsson, K. and Schmid, H. (2001), 'Real Integration or Just Formal Adaptation?', Paper presented at *the Conference of the European Sociological Association*, Helsinki, p.17.

75 Jacobsson and Schmid (2001:17).

but of voters in the two regions. Therefore, a more likely explanation is that the similarities may reflect both a wider political change and a change in how regional development is thought of and explained in academic circles. Regional representatives are in this context to be regarded not as recipients of ideas and beliefs but as co-producers of them.

A Cognitive Shift in the Regional Debate

During the post-war period and up to the early 1980s, western European regions were not primarily understood as competitors but as complementary parts in a national totality.[76] During this era, regional policy-making tended to proceed from the idea that the central state both *could* and *should* use a wide range of interventionist instruments for influencing the spatial distribution of economic activity.[77] But this 'old' model of development, sometimes referred to as 'spatial Keynesianism',[78] came to be widely called in question in the 1980s. Economic development strategies ceased to focus on redistribution and equality, and came to engage with attempts to create endogenous growth in all regions.[79] The tendency to stress regional competitiveness and endogenous potentials rather than complementarity and redistribution has previously been understood as a cognitive shift in the regional debate.[80]

This shift of focus has been explained by references to the economic crisis and the subsequent reduction of public expenditure that took place in many western European states in the late 1980s and early 1990s. Economic decline made the central state's position as the redistributor of wealth and welfare between regions more uncertain, and regions were actually compelled to solve their structural and economical problems increasingly on their own.[81] One argument was that globalisation gradually diminished the state's ability to regulate economic interests, and that the emergence of new international political organizations, new methods of communications, and an increasing amount of multi-level governance within the EU made the state more or less redundant.[82] The shift in

76 Keating (2001:217); Keating *et al.* (2003:6,18).
77 Keating (1998:140).
78 Keating *et al.* (2003:15).
79 Deas and Ward (2000:274).
80 Lovering (1999:380); Keating *et al.* (2003:11).
81 Tomaney and Ward (2000).
82 for a discussion on that, see Jones (2001:1188); Deas and Ward (2000:276); Lash and Urry (2002:280); Gruber, Karoly (2001), 'Regionalism, Nationalism, Integra-

the regional development debate did take place contemporaneously with EU initiatives to improve the economic position of the community as a whole. However, this should not be taken as evidence of the EU's one-sided ascendancy at the regional level. Thus, the regional level has not only been a recipient of policy but has also proved to be increasingly engaged with 'the institutions and processes of EU-policy making'.[83] The tendency to stress competitiveness should thus not be regarded as a result of top-down EU policy only, but as as a result of collaboration with the member states and with representatives for the regional levels.[84]

The shift of focus from redistribution to endogenous growth was, as indicated above, stimulated by political actors at all levels. Yet arguments about the changing capacity of the state and about the increasing importance of the regional level were advanced also in academic circles.[85] Hence, some of the critique against the interventionist state was voiced by social scientists and economists within the universities.[86] A state that is too caring, too demanding or just too present was thus said to suffocate all sorts of initiatives at a local level. 'Numerous and extensive programmes for welfare' were accused of having a weakening effect on non-governmental organizations, and of making informal social relations unnecessary.[87] Fukuyama – a well-known critic of the state – even suggested that the mere intervention of the state might weaken the society's 'underlying propensity for spontaneous sociability in the long run'.[88] During this

---

tion; Central and Western European Perspectives', *The Challenge Europe On-Line Journal 17/1*; Kohler-Koch, Beate (2002), 'European Networks and Ideas: Changing National Policies?', *EioP*, 6., p.1.

83 Jeffery, Charlie (2000), 'Sub-National Mobilization and European Integration: Does it Make any Difference?', *Journal of Common Market Studies* 38 (1), pp.1-23., p.1; Marks, G., Nielsen, F., Ray, L. and Salk, J. (1996), 'Competencies, Cracks and Conflicts', in G. Marks, F.W. Scharpf, P. C. Schmitter, and W. Streeck (eds.), *Governance in the European Union*, London: SAGE, p.61.

84 Keating *et al.* (2003:18).

85 Lash and Urry (2002:279); Hallin, G. and Malmberg, A. (1996), *Attraktion, konkurrens och regional dynamik i Europa*, Stockholm: Fritzes, p.15,

86 See Florida, Richard (1995), 'Toward the Learning Region', *Futures*, 27(5), pp.527-536, p.528; Giddens, Anthony (1999). *Tredje Vägen – om förnyelsen av socialdemokratin*, Stockholm: Atlas.

87 Rothstein, Bo (2000), 'Socialt Kapital i den socialdemokratiska staten. Den svenska modellen och det civila samhället', *Arkiv* nr 76, p.1.

88 Fukuyama, Francis (1995), 'Social Capital and the Global Economy', *Foreign Affairs*, 74(5), pp.89-103, p.89.

period, the saying that the nation-state is 'too small for the big problems and too big for the small ones' turned into an academic and political catchphrase.[89]

Some of the scholarly critique against the state and against top-down regional policy measures was based on the assumption that the industrial era was long past, and that we were about to enter an 'age of information', 'a new age of capitalism', or a 'knowledge society'.[90] Given the alleged shift to a 'knowledge-driven economy',[91] economic success was no longer thought of as achievable by means of state subsidies or by support to large-scale industries on the periphery. If anything, regional economic development was understood as resting upon a set of 'non-state and non-economic factors such as sub-national social, cultural and institutional forms'.[92] This assumption (to all appearances made also within the regions studied here) implied that a mere redistribution of wealth would be untenable in the long run. Instead, regional policy measures should aim at fostering institutions favourable to economic growth *within* the regions.[93]

Gradually, it became commonplace to try to explain variations in economic strength, not by reference to differences in access, supply of raw materials or fixed assets, but by reference to the so-called soft location factors.[94] A number of scholars have stressed the importance of 'civic spirit'[95] or 'institutional thickness'.[96] Others have brought out the region's 'technological capacity' or access to 'know-how' as essential factors for the economic development.[97] Yet

---

89 Bell, Daniel (1973), *The Coming of Post-Industrial Society: A Venture in Social Forecasting*. New York: Basic Books, quoted in Lash and Urry (2002:279); Williams, Raymond (1983), *Keywords*, New York: Oxford University Press, p. 197-9; Jönsson, C., Jerneck, M., and Stenelo, L. (2001), *Politik i globaliseringens tid*, Lund: Studentlitteratur, p.18.

90 Gertler, Meric, S. (2003), 'Tacit knowledge and the economic geography of context or The undefinable tacitness of being (there)', *Journal of Economic Geography*, 3, pp.75-99, p.75; Florida (1995:527).

91 MacKinnon, D., Cumbers, A. and Chapman, K. (2002), 'Learning, innovation and regional development: a critical appraisal of recent debates', *Progress in Human Geography* 23(3), pp.293-311, p.294 .

92 Jones (2001:1186).

93 Deas and Ward (2000:274).

94 Törnqvist (1998:67ff.); Keating *et al.* (2003:8)

95 Putnam, Robert D. (1993), *Making Democracy Work – Civic traditions in modern Italy*, Princeton, NJ: Princeton University Press.

96 MacKinnon *et al.* (2002:294).

97 Florida (1995:528).

others have argued that creativity, the ability to learn, the capacity to innovate[98] or the region's capacity for flexible specialization[99] are of vital importance for a region's competiteveness. Others again have talked about 'soft location factors' in more general terms.[100] All in all, we meet a number of scholarly voices arguing that the regional software is far more important for the regional economy than the prospects of obtaining subsidies from the central state. It is not the 'inherited (dis)advantages'[101] that are in focus, but the 'constructed advantages'.[102] In economic geography, the shift of focus towards the effect of 'habituated routines and conventions' on economic activities has been interpreted as an 'institutional turn'.[103] In related disciplines, it has been seen as part of a 'new regional development paradigm'.[104] This new thinking, it has been argued, has had an 'immense impact on regional development policies across Europe and within the European Commission itself'.[105]

A New Spatial Orthodoxy

So far, I have argued that the discursive similarities across the regions and across party political and organizational borders should be seen as a result of a cognitive shift in the regional debate. The word-view that comes up in the two regions is thus not an effect of specific, regional circumstances. Nor are the notions about globalization, Europeanization and competitiveness to be regarded as a simple, one-sided adoption of 'eurojagon'.

Another and more central argument, however, is that the world-view presented above forms part of a regionalist ideology and contributes to a process of region

98 Törnqvist (1998:120ff); Cooke, Phil (1997) 'Institutional Reflexivity and the Rise of the Region State' in Benko, G. and Strohmayer U. (eds), *Space & Social Theory: Interpreting Modernity and Postmodernity*, London: Blackwell.

99 MacLeod, Gordon (2000), 'The Learning Region in an age of austerity: capitalising on knowledge, entrepreneurialism, and reflexive capitalism', *Geoforum* 31(2) pp.219-236, p.219

100 Ceccato and Persson (1999).

101 Keating *et al.* (2003:8).

102 Cooke, Phil and Leydesdorff, Loet (2005), 'Regional Development in the Knowledge-Based Economy: The Construction of Advantages', *Journal of Technology Transfer*, 30(3), forthcoming.

103 Amin (2001:1237). For an informative account for different strands in economic theory, see MacKinnon *et al.* (2002:295-297); see also Jones (2001:1188)

104 Keating *et al.* (2003:14).

105 Keating *et al.* (2003:18).

building. Thus, in *nationalist* ideology, a world consisting of *nations* is seen as the natural, the desired or perhaps the only possible world.[106] Nationalist ideologies tend – whether they have ethnic or civic, 'hot' or 'banal' overtones – to treat the nation as the given and most natural basis for social and political action.[107] Besides, even those who would never label themselves as nationalists did act within a cognitive framework where nations and nation-states appeared as the 'natural' and given basis for political action.[108] The way in which regional development issues were handled in the recent past is actually a very illustrative example. The 'old' way to approach regions was to conceive of them as parts of a national totality. Top-down redistributive regional policy is, then, both an example of a policy that contributed to *maintaining* the idea of the nation state as the given political entity, and a *result* of this very idea.

These empirical and theoretical expositions, however, indicate that the 'state-centred ontology' is on the way out, or at least is about to lose its role as the predominant geographical paradigm.[109] This ontology is challenged and replaced by what Jones and others have called a 'new spatial orthodoxy', supported not only by policy makers but also by a wide range of academics.[110] This new orthodoxy rests, according to Lovering, on the assumption that '"the region" is becoming the "crucible" of economic development … and on the normative bias that "the region" should be the prime focus of economic policy'.[111] 'New economic rules' are said to make regions replace the nation state as the 'fundamental building blocks or motors' of current political and economic life.[112] Academics as well as policymakers appear to understand the regional – and not the national – scale as the scale at which 'a whole series of economic, political, and cultural contexts' should be observed.[113] To judge from the empirical section above, it is certainly not a world of *nations* but a world of *regions* that is perceived as natural or desirable.

---

106 Billig (1995:37).

107 Kellas, James G (1998), *The Politics of Nationalism and Ethnicity*, London: Macmillan Press, p.28; Billig (1995).

108 Löfgren, Orvar (1989), 'The Nationalization of Culture', *Ethnologia Europaea* XIX, pp.2-23, p.9.

109 MacLeod and Jones (2001:672, 679).

110 Jones (2001:1185).

111 Lovering (1999:380); see also Jones (2001).

112 Berggren, Christian and Brulin, Göran (2002), *Klistriga kluster eller globala glidare. Den lokala dynamikens paradoxala betydelse i den globaliserade ekonomin*, Stockholm: NUTEK; Florida (1995:528); For a critical discussion, see Lovering (1999:383); Paasi (2000).

113 MacLeod and Jones (2001:671).

Regional Self-Centrism

The increasing focus on regions implies an increasing need for a definition of the region concept, or at least a discussion of 'the crucial question of what a region is'.[114] Nonetheless, there is a tendency, in contemporary academic discourse as well as in political strategy, to take the 'foundational concept of "the region" for granted' or to 'script the region as a pre-given boundary'.[115]

The critique of the tendency to take the region for granted is very convincing. The region concept is certainly used in a multitude of contexts without a vestige of a definition. It is true that the conviction that every region should be transformed into a 'learning', 'intelligent', 'innovative' or 'competitive' space fails to consider the ambiguity of the region concept.[116] Regions are territorial units that always exist in relation to some particular criteria.[117] Particularly when politicians wish to transform 'their' political region into a 'learning' or 'innovative' one, they completely ignore the fact that the boundaries of 'learning' or 'innovativeness' are very unlikely to match the boundaries of the formal, political region.

Yet this is an approach to regions that is wide open to introspective discussions about the *content* of the region. Hence, an adjacent argument has been that the 'new age of capitalism' requires 'a new kind of region'.[118] For that reason, we can observe a growing tendency in academia as well as in politics to analyse and to dissect the region and its constituent parts. In accordance with this development, *place* has come to play an increasing role in academic thinking.[119] The focus on endogenous potentials and on their alleged importance for economic development has in fact resulted in countless attempts to make an inventory of the human resources, social capital, innovative milieus or other so-called soft factors in the region.[120] So the *boundaries* of the region may be taken for granted, but engagement with its *content* is greater than ever.

114 Paasi (2000).

115 MacKinnon *et al.* (2002:297); MacLeod and Jones (2001:670); Jones (2001:1202).

116 For a discussion on 'new' kinds of regions, see Cole and Loughlin (2003:266); Lovering (1999:388); Jones (2001:1190); Florida (1995:532); Berggren and Brulin (2002).

117 Keating (1997:17); MacLeod (2001:811); Keating (1998); Allen *et al.* (1998:1).

118 Florida (1995:532).

119 Keating (2001:218).

120 MacKinnon *et al.* (2002:294).

Another problem with the increasing focus on regions is the tendency to approach regions as if they were *actors in their own right.*[121] In this topic, Paasi has argued that it is 'still quite common to understand regions as active "actors" ... capable of doing or changing things'.[122] Harvey raises a similar objection in a discussion about cities, pointing out that there is a tendency to regard cities as active agents when they in reality are 'mere things'.[123] To talk about regions as actors is, as MacLeod and Jones argue, to 'fetish space and to reify places as if they themselves – rather than classes or regional alliances are the active agents'.[124] In the regions studied here, we can easily observe a tendency to group all kinds of actors, institutions and localities together and to understand them as parts of a regional whole. The regions are almost portrayed as organisms, exposed to the same competitive pressures and capable of striving in one direction.[125] On this topic, Lovering has argued that speaking of competitiveness 'as an imperative at the level of a region – rather than a firm – is almost completely uninformative'.[126] Hence, it is argued, it is *firms* that are subjects of competition and *firms* that succeed or fail in global competition – not territorial units. Still, the tendency to see the region as a coherent whole faced with increasing competition has created a demand, not only for *new* concepts and models of development as MacKinnon *et al.* suggest, but for *regional* concepts and models as such.[127]

A Basis for Political Action

It is open to discussion whether the spatial orthodoxy referred to above should be understood as a real and substantial change, or whether, as both Jones and Lovering claim, we are facing yet another trend that overemphasizes or 'talks up' scale in order to 'justify a particular theoretical perspective and policy narrative'.[128] It could also be questioned whether the state really has retreated or whether, as Keating suggests, it has rather been 'penetrated by new influences...

---

121 Keating (1997:39).

122 Paasi (2000).

123 Harvey, David (1989), 'From Managerialism to Entrepreneurialism: The Transformation in Urban Governance in Late Capitalism', *Geografiska Annaler B*, 71(1), pp.3-17, p.5.

124 MacLeod and Jones (2001:670).

125 See MacKinnon *et al.* (2002:296).

126 Lovering (1999:388-9).

127 MacKinnon *et al.* (2002:296).

128 Jones (2001:1189); Lovering (1999:388-9).

(and) complex patterns of negotiation'.[129] Indeed, many of the issues raised above as 'new' could be understood as old ideas in new packaging. The 'regional question' certainly did exist prior to the aforementioned discourses on globalization and the retreat of the state. Besides, the highlighting of endogenous regional growth capacities surely represents, as MacKinnon and others have pointed out, an 'important continuity with earlier work on new industrial districts and spaces'.[130]

Whether or not the current spatial orthodoxy has any bearing in 'reality', and whether it is new or long-established, there is a strong consensus around several notions within the two regions studied. There is, to begin with, a consensus around the increasing importance of regions. There is, furthermore, a consensus around the importance of regional competitiveness. This, in turn, is followed by widespread approval of the existence of some soft location or development factors that would increase the regions' ability to cope with global competition. These assumptions, I would say, constitute the essence of the world-view of the kind of regionalism we encounter in Norrbotten and Mecklenburg–Western Pomerania. The question now is what repercussion this world-view is having on political action in the two regions. How does the conception of the world as globalized, regionalized, and competitive relate to the core values communicated by regional representatives?

---

129 Keating (1997:27).

130 MacKinnon *et al.* (2002:296); see also Cole and Loughlin (2003:266); Keating *et al.* (2003:14f).

5. The Core Values of Regionalism

By now, we have learned that much of the political debate in both Norrbotten and Mecklenburg–Western Pomerania proceeds from the belief that the state is on decline and that the regional level is increasingly important. This is, as it seems, well received, and few objections are made to the alleged rise of a 'Europe of the Regions'. This very point of departure – that regions are widely recognized as important – is very different from the point from which sub-national or regionalist movements during the 1960s and 70s started to make their claims. Many of the regionalist movements that were prominent in European politics during that period have been associated with a problematic struggle for autonomy, for the 'right to roots' or for cultural recognition.[1] In the regions studied here however, neither cultural recognition nor political autonomy seems to be the main issue. European integration is understood as a favourable turn for the regional level, and the supremacy of the nation state is considered to be on the way out. The 'new regionalist' agenda, on the other hand, has been characterized by its claims for increased economic autonomy – claims from which less favoured regions like Mecklenburg–Western Pomerania and Norrbotten would not necessarily gain.

Yet most ideologies tend to defend some certain values or interests. The question we need to raise, then, is what political agenda the regional representatives seem to follow. What are the values emphasized in the regional policy discourse? How do regional policy-makers understand the 'interests of the region's population',[2] and what would a 'good society' look like, according to the same group?

NORRBOTTEN

Competitiveness and Growth

In many of the policy documents analysed here, 'economic growth and development' stands out as the overriding political goal. We find calls for social,

1 Gren (1999:27,32).
2 Williams (1997:114).

humanitarian, environmental and other goals too,[3] but strengthening the competitiveness of the region and supporting long-term economic growth stands out as an agenda that almost all actors within the region endorse.[4] There are, indeed, those who call for subsidies from the state and from the European Union. In general, however, dependency on the central state is depicted as something truly problematic. Most actors seem to have adopted the idea that every region should provide for its own economy, and accordingly economic independence rather than subsidies seems to be the chief political goal.[5]

The vision of how this economic independence is to be achieved is indeed somewhat scattered. Control over the profits from the water-power plants and from the mining and forest industries in the region has been a recurrent topic in the Norrbottnic political debate over the years. Even today we find arguments that the region is exploited by the central state, and that a regionalization of large-scale industries and natural resources should be the main political goal.[6] Far more common, however, is the idea that it is regional initiatives and small and medium-size companies that will get the Norrbottnic economy going. But, it is argued, there is a need for a social climate in which these initiatives and these companies can come forward. Given that, 'unity', 'cooperation' and 'joint efforts' turn out to be examples of highly praised 'values' in the region.

3 SAP Norrbotten (2001); SAP Norrbotten (2002), 'Tillsammans – för en bättre framtid i Norrbotten', *Regionalt Valmanifest*, Norrbotten; p,1; NLL (1997c:11); NLL (1998c:13).

4 Norrbottengruppen (1999); NLL (1997c:11,21), NLL (1998c:13); NLL (1999:14); NLL (1997a:5); NLL (1996a) *Landstingsplan 1997*, Luleå, p.12; NLL (2000b) *Landstingsplan 2001*, Luleå, p.11; SAP Norrbotten (1997:6); SAP Norrbotten (2002); Folkpartiet Liberalerna Norrbotten (2002b), *Vi tycker att du ska rösta på oss!*, Luleå, p.2; Folkpartiet Liberalerna Norrbotten (2002a:5); Törnman, Lars and Backgård, Kenneth (2002), 'Norrbotten en tillväxtzon', *NSD*, 2002-06-10; Westman, Bengt and Sundström, Anders (2002), 'Tornedalen kan bli en ny tillväxtregion', *NSD*, 2002-09-14; Zakrisson, Kristina (2000), 'Varje region behöver en egen tillväxtpolitik', *NSD*, 2000-09-28; Zakrisson (2003); SAP Norrbotten (2004a) 'De här åtgärderna vill vi se', *Uttalande från DS* 2004-02-03.

5 For an alternative understanding, see Abrahamsson (2002).

6 See Centerpartiet Norrbotten (1998:10); Vänsterpartiet Norrbotten (1998:27); Norrbottens Sjukvårdsparti (1998); Sundkvist, Hans-Åke (2003), 'Vi vill inte ha bidrag, bara vår beskärda del', *NSD*, 2003-05-30.

Common Goals and Strategies

A wealth of examples of calls for mobilization and unity can be found in *The Guide*, referred to in Chapter 4 as a document that proved to be central to the political debate in the county during the 1990s. When *The Guide* was made public, the broad formal support that the document enjoyed was cited as one of its main advantages. Some voices even held that the cooperation that surrounded the making of the document was far more important than the document itself.

Within *The Guide*, political actors and geographical units within the region are asked to bring their conflicts to an end and to replace them with cooperation.[7] The alleged need for unity is occasionally illustrated with metaphors. We learn that 'unity is strength'[8], that it is time to 'bury the hatchet'[9] and to 'smoke the pipe of peace'.[10] It is obvious that the level at which this peaceful cooperation should take place is the regional one. A deep, popular engagement in ones' own village or ones' own municipality is indeed considered both positive and desirable. Nevertheless, it is emphasized that the well-being of single localities should by no means be allowed to jeopardize the well-being of the region as a whole. Local engagement is good, it is argued, but parochialism[11] is referred to as something that on no condition should get the upper hand in the regional developmental debate. According to *The Guide*, there is a need for an attitude that 'allows all of us to value the region's best interest more than we value the interests of ones' own municipality or village'. In developing the region, it is argued, 'we are not village citizens, Saamis, men, women, enterprisers, public employees, farmers, industrial workers, party members, civil servants or immigrants – but Norrbothnians'. 'Now if ever', it is declared, there is need for 'a strong Norrbotten spirit'.[12]

The idea that there is a need for regional unity and a 'Norrbotten spirit' is linked to the idea of the region as a competitor in a global race for economic growth and development. Many voices dissociate themselves from the old top-down regional policy and from dependence on state subsidies. In this respect, *The Guide* maintains that Norrbottnic history has been coloured by a lack of agreement and an insufficient will to cooperate. Previous top-down regional development policy measures, according to *The Guide*, have resulted in 'proper

---

7 Partnerskap för Norrbotten (1995:10-12).

8 *Ensam är inte stark* in the original.

9 *Gräva ner stridsyxan* in the original.

10 *Tända fredspiporna* in the original.

11 *Byatänkande* in the original.

12 Partnerskap för Norrbotten (1995:10,12,18).

scrambles' in the region, where 'everyone tried to grab what they could in terms of investments or subsidies'. These scrambles are said to have caused a split between actors within the region, and to have resulted in a situation where 'everybody lost from the bargain'. Today, it is argued, these conditions are about to change and 'increasing harmony' is said to be emerging 'between municipalities and authorities in our county'. This alleged harmony is brought to the fore as an important resource for regional policy-makers to make the most of. 'Yet', it is acknowledged, 'we need to overcome differences and to further improve our ability to cooperate.' It is also contended that in the present situation 'we cannot afford to fight over the establishment of companies, the erection of hospitals or over the launching of tourist campaigns'. We need, it is argued, to get 'better at finding solutions of advantage for the whole county and at rejecting those suggestions that impair our ability to compete with other regions'. Consensual cooperation is important, it is suggested, since 'we cannot afford to let disagreements within the county weaken our competitiveness'. 'We cannot act alone', it is argued. 'Individuals, villages, municipalities, authorities, companies, organizations, schools and others need to stick together and cooperate in order to achieve results.'[13]

Concord, Spirit and Will

Even if *The Guide* proves to be an early and extraordinary striking example of how calls for concord and regional unity could be framed, the message in that document is rather commonplace. When the possibilities of establishing a regional body for self-government in Norrbotten were under debate, the Regional Association of Local Authorities (RALA) and the County Council produced several investigatory reports in which 'concordance and united action' or 'cooperation and a distinct political will' were brought out as essential for regional economic development.[14] Some of the calls for institutional reform were indeed based on the argument that decentralization of political power from the state to the region would enhance democracy. In these cases, it was argued that power over the region 'needs to start out from the Norrbothnians themselves'.[15] Another and more common line of argument, however, was that a change of the institutional order, and increased power over regional development policy, would

---

13 Partnerskap för Norrbotten (1995:8,18).

14 Regionutredningen Norrbotten (1999a); Regionutredningen Norrbotten (2000b).

15 Regionutredningen Norrbotten (1999b:8).

augment the chances for economic development in the region.[16] This rationale was based on the notion that a decentralization of power from the state to the region would stimulate political activity in the region. This political activity would then foster a civic spirit and thereby stimulate economic activity. This notion was in some cases qualified with the argument that 'researchers have concluded that economy is to a great extent determined by the civic spirit'.[17]

The value of concord, unity and a certain regional spirit is emphasized in other contexts, too.[18] 'Constructive and cooperative politicians' are referred to as a necessity for the region, and representatives from several parties stress the importance of 'teamwork' and mutual understanding in the region.[19] In *The Guide*, we met with the proverbial saying 'unity is strength'. Later on, this phrase has been repeated also by Social Democratic politicians.[20] A representative for the Health Care Party considers the capacity to cross party frontiers, to undertake joint action, and to advocate a common political position as an end in itself and as a means of bringing economic development to the region.[21] A number of actors stress that 'cooperation, rather than confrontation', 'a genuine feeling of fellowship', 'broad cooperation between municipalities', and 'distinct Norrbottnic positions and points of view' are intrinsic values, as well as a paths to regional economic development.[22]

The County Council's Media Policy Programme emphasizes that 'we need to stick together and to develop a firm Norrbotten spirit in order to handle the challenges of the future'.[23] In the Action Plan (AAP) from 1999, the 'capacity to

---

16 Regionutredningen Norrbotten (2000b:9).

17 Regionutredningen Norrbotten (1999b:10), (2000a:9).

18 Marklund, Kari (2000) *Redovisning av regionutvecklingsdagen på Pite havsbad 00-03-01*, Luleå: Lst Norrbotten.

19 Gröning-Degerlund (2003), 'Dags för enighet och strategi', *NSD*, 2003-08-18; Hedberg, Göran (1998), 'Politik är ett Lagspel', *Hjärtslag*, Nr 5; Ström, Mårten (2002), 'Luleå mot en ljus framtid', *NSD*, 2002-05-31.

20 Westman and Sundström (2002).

21 Backgård, Kenneth (1998b), 'Samarbete – Fördel Boden', *Hjärtslag*, Nr 4, p.14.

22 Westman, Bengt (2002), 'Fler regionalpolitiker behövs', *NSD*, 2002-06-14; Fredriksson, Sture and Niva, Sören (2002), 'Samarbete istället för konfrontation', *NSD*, 2002-09-04; Sundström, Anders (2003), 'Socialdemokraterna har ett tungt ansvar för länet', *NSD*, 2003-03-21; Eriksson, Stig and Haarala, Ove (2003), 'Politik för hela länet – mer än bara sin egen bakgård', *NSD* 2003-07-28; Eriksson, P.O (2003b), 'Företagandets, samarbetets och möjligheternas län', *NSD* 2003-09-11; Norrbottengruppen (1998); Lst Norrbotten (2001b) *Internationellt Handlingsprogram*, Luleå.

23 NLL (1996c:7).

mobilize the regions' own resources' is understood as an 'ever more important factor for regional development'.[24] In the AAP from the year 2000, 'the 'Norrbothnians' prospects of a good life in the region' are said to depend on the 'extent to which regional actors are capable of making a real, joint effort to mobilise the resources of the region'.[25] When the resources are specified, a wide range of private and public institutions in the region is said to be capable of strengthening the competitiveness and the development possibilities of the region.[26] A year later, the County Governor stresses in a signed article that social capital and civic spirit are important resources for the development of the entire region. Referring to research by Robert Putnam, the County Governor suggests that social capital in Norrbotten is a joint resource that must be used if the county is to keep up with the international competition.[27] Thus, it is argued, 'United we stand, Divided we fall'.[28]

In the Regional Growth Agreement (RGA), agreed to by a wide partnership of actors in 1999, we learn that the drawing up of a common developmental plan and unification around a common analysis and a common cause have been of outmost importance for the region.[29] In the subsequent Regional Growth Programme (RGP), it is stressed that unity and common positions in regional development issues would benefit the region. The work of bringing the region together is regarded a priority for action, and according to the RGP 'all actors within the region should be able to ... see the long-term advantages of a common focus ... even if the own organization do not benefit from it at first sight'.[30]

Regional development policy is but one policy field in which calls for concord and unity are evident. The value of unity comes to the fore also when international affairs are under discussion. Hence, the desire to play an active role in the EU policy process has contributed to the demand for 'regional views' and common regional stances in many regions.[31] In the 1998 International Policy

---

24 NLL (1998c:6).

25 NLL (1999:5).

26 According to the County Council, medical service, associations, schools, companies, culture centres are , to be seen as resources in this respect . NLL (1999:16).

27 Marklund, K. (2001b), 'Viktigt att förvalta Norrbottens sociala kapital', *NSD*, 2001-12-12. See also Putnam (1993).

28 Marklund (2001b).

29 Norrbottengruppen (1999:6).

30 Lst Norrbotten (2003c), *Norrbottens tillväxtprogram 2004-2007*, Luleå.

31 Kommunförbundet Sörmland (1999), *Strategi för länsövergripande samverkan i Internationella frågor*, Sörmland; Västmanland International Co-operation (2001) *Internationell strategi och handlingsplan för Västmanlands internationella nätverk (VIN)*, Västerås.

Programme (IAP) for Norrbotten, the then County Governor stresses that dependency on the surrounding world requires a common position within the region.[32] The subsequent County Governor argues in a later IAP that the new world order demands a proactive attitude and work to promote Norrbottnic positions and interests.[33] In an article on the same topic, the County Governor in question claims that Norrbotten needs to 'present a united, well-ordered front' towards the commission and Swedish representatives in Brussels.[34] Overall, regional unity, mobilization and 'spirit' stand out as values promoted by many actors within the region.[35] We do indeed find those who stress differences in conditions between different groups and locations in the county.[36] But even when the differences are acknowledged, it is stressed that development in Norrbotten 'stands and falls with cooperation'.[37] Close cooperation around a common cause and a strong engagement in the county's common interests are thus values that actors from most political camps seem to unite around.[38]

Mecklenburg–Western Pomerania

Unity and Growth

While one analyses political documents in Mecklenburg–Western Pomerania, it is evident that a weak economy and high unemployment are perceived as the main problems for the *Land*. Accordingly, the 'better society' longed for in MV is one in which the *Land* has improved its competitiveness and growth rate, where fewer people are excluded from the labour market, and where the *Land* has become more attractive for foreign investors.[39] It is, to some extent, maintained that there are groups in MV that are particularly vulnerable, and therefore

32 Norrbottengruppen (1998:3).

33 Lst Norrbotten (2001b).

34 Marklund, Kari (2003), 'Nio miljarder står på spel våren 2003', *NSD*, 2003-03-13.

35 See also Hedberg (1998:2).

36 SAP Norrbotten (2001:3); Norrbottengruppen (1999:12).

37 SAP Norrbotten (2001:3).

38 Eriksson, S. *et al.* (2003); NSD (2002a); Vänsterpartiet Norrbotten (1998:4); NLL (1999:16); NIR (2003b).

39 StK MV (2002a); StK MV (2002f), *Regierungserklärung zum Beginn der 4. Legislaturperiode*, Schwerin; StK MV (2002d), 'Schwerpunkte der Landesregierung: Standortoffensive und Wirtschaftsförderung', *Rede anlässlich der Eröffnung des 3. Fördertages der Unternehmensverbände*, Schwerin, p. 2; CDU Landtagsfraktion MV (2002:8); RPV Vorpommern (1998:14).

in particular need for support.[40] Still, unity, cooperation and joint action are held in high esteem, just as they are in Norrbotten . There also seems to be faith in a policy that would benefit everyone in the *Land*.

In Mecklenburg–Western Pomerania, calls for unity and consensus have a somewhat different ring from those in Norrbotten. The GDR past and the fact that MV was the first *Land* in Germany in which the Social Democratic Party (SPD) formed a government with the Party of Democratic Socialism (PDS) may have combined to the many calls for reconciliation and joint efforts.

In the 1998 coalition government's parliamentary address, we learn that it is necessary to 'bring our divided society together' and to surmount the barriers between people in the *Land*. In the address, the alleged gulfs between 'victims and perpetrators during the GDR times', between 'those who benefited from *Die Wende* and those who lost out', between 'the right and the left', and between 'East and West Germans' are referred to as problems that the region needs to overcome in order to develop. Progress, it is argued, must be accompanied by solidarity and cooperation, and the ambition to overcome differences within the *Land* is therefore put forward as an action priority for the new government. The SPD President expresses, furthermore, a desire to establish 'a new attitude' between the parties in the government and between the government and the opposition. This new attitude should be 'straightforward in factual matters, fair in manner and always work for the *Land*'s best interests'. The parliamentary address also suggests that, without reconciliation with the past, it will be impossible to create a positive image of the future. The strength needed for reconstructing the *Land* after *die Wende* will, according to the government, 'not be won by splitting and polarization' but will be achieved 'only if we unite and make these efforts together'. To judge from the parliamentary address, no one should be excluded from the struggle for the common good of the *Land*. Citizens, industry, trade unions, experts or 'everyone with good intentions' are to be invited to the work on reconstructing the *Land*.[41]

Yet the address from 1998 is not unique when it comes to calls for unity and joint effort.[42] Two years later, the invitation to work for the common good of the *Land* is repeated in a government report. 'Help us with our efforts', it is announced, 'let's work jointly for the development of our *Land*'. Thereafter, 'everyone who with their soul, heart and good ideas wants to contribute to the

40 StK MV (1998:14).

41 StK MV (1998:3-6,14).

42 StK MV (2000b), *Rede anlässlich des 10-jährigen Bestehens des Bundeslandes Mecklenburg-Vorpommern*, 2000-10-18, Schwerin; StK MV (2000c) *Entwicklungsbericht Vorpommern*, 2000-11-23, Schwerin, p. 104; Sinz (2000:9).

development of the *Land*' is invited to undertake developmental work. [43] In a speech of around the same time, the President announces that the *Land* has applied to host a BMW factory, and that it does so 'together with more than 100 other applicants worldwide'. In this context, it is argued that MV can offer 'good and suitable locations in Mecklenburg–Western Pomerania', but that this is far from enough. In order to succeed with the application, it is contended, 'we ... have to make clear that all political forces in the *Land* stand behind this application'. When applying for projects like this one, it is argued, it is crucial that 'all stand together'. Therefore, the President emphasizes that he 'appeals urgently' to the audiences' 'sense of responsibility' and asks them to note that every '*parteipolitisches Taktieren*[44] would 'harm our *Land*' and 'impair our chances to succeed with the application'.[45]

In the parliamentary address from 2002, greetings are once again sent to 'all those who have participated in the struggles for the *Land*'. All sorts of actors are mentioned as co-workers in the efforts to bringing the *Land* forward. Members of parliament, representatives of the economy, the trade unions and the churches, media, sports movements, the army and 'committed individuals' all receive expressions of gratitude in the report.[46]

Common Goals and Strategies

In the parliamentary address from 1998, the work on improving the economy and on lowering unemployment rates is said to be unfeasible unless all actors unite in their efforts and start 'pulling together'.[47] The importance of acting in concert or of 'pulling together' reappears in other contexts too. It comes up in the coalition government's half-time report from the year of 2000, in discussions on how to promote MV as an attractive tourist resort, in dialogues on vocational training, and when the economy in general is referred to.[48] In the report from the year

43 StK MV (2000a) *Es geht voran in Mecklenburg-Vorpommern*, Schwerin.

44 'Party-political tactics'.

45 StK MV (2000b).

46 StK MV (2002f).

47 *An einem Strang ziehen* in the original. Literally, 'pulling the same rope'. StK MV (1998:10).

48 Finansministerium MV (2000), 'Einbringung des Haushaltsplan-Entwurfs 2001', *Pressemitteilung Nr 60/00,* Schwerin; MfABL MV (2000f), 'Vorpommern - Landesteil der besonderen Chancen', *Pressemitteilung Nr 99/00*, Schwerin; MfWi MV (2004), 'Starker Auftritt und klare Botschaft - MV tut gut', *Pressemitteilung Nr 42/04*, Schwerin; StK MV (2000a:18); StK MV (2001), 'Bekämpfung der Arbeits-

2000, it is held that 'we in Mecklenburg–Western Pomerania will only be able to bring our *Land* forward jointly' and in the 2002 parliamentary address the value of 'pulling together' is once again stressed.[49] In other contexts, it is not the importance of 'pulling together' that is referred to, but the import of following a common *Leitbild*. The PDS Minister for Regional Development contends in this matter that the prerequisites for development would improve if all relevant actors within the region could rally around a common *Leitbild*. This *Leitbild*, it is furthermore contended, can only be 'successfully implemented when all are steering in the same direction'.[50]

As in Norrbotten, unity and joint efforts are seen as important also in external relations. In this matter, the President stresses that 'we need to bring forth the *advantages* of our region jointly, rather than to slander ourselves'.[51] The President, as well as other actors, suggests that citizens, political representatives and private entrepreneurs should act as 'ambassadors' for the region, putting forward its positive sides rather than its drawbacks.[52] In a later speech on the same topic, the President again points out that 'some are experts in telling the surrounding world what's wrong with MV' and that this slandering needs to be replaced by more positive narratives from the region.[53]

Voices from the Sub-Regions and the Opposition

So far, we have seen a number of examples of the SPD/PDS coalition government calling for unity, cooperation and reconciliation. One may indeed suspect

---

losigkeit ist größte gemeinsame Aufgabe', *Pressemitteilung Nr 93/01,* Schwerin; StK MV (2003c), 'Gemeinsam an einem Strang ziehen und Land weiter voranbringen', *Pressemitteilung Nr 137/03,* Schwerin; PDS Landtagsfraktion MV (2003), 'Gemeinsam die Ausbildung von Jugendlichen sichern', *Presseerklärung*, 2003-05-28, Schwerin; Landtag MV (2004b), 'Landtag MV tritt dem Olymp-Club bei', *Pressemitteilung*, 2004-03-25; Ebenet, Otto (2004), 'Tourismuskonzeption für Mecklenburg-Vorpommern', at *www.investorenportal-mv.de/archiv/211.html*, 29. Juni 2004.

49 StK MV (2000a:18); StK MV (2002f) .

50 MfABL MV (2003b), 'Start von Aktionsprogramm für Firmen, die neue Wege gehen', *Rede des Arbeitsminister Holter auf Regionalkonferenz Mecklenburgische Seenplatte*, 2003-04-24, Schwerin.

51 StK MV (2002a).

52 StK MV (2002a); StK MV (2002f); LandtagsNachrichten MV (2002), *Quoting Bretschneider, Sylvia*, 2/2002, Schwerin, p.16; LandtagsNachrichten MV (2003b), *Quoting Ringstorff, Harald*, 1/2003, Schwerin, p.16.

53 StK MV (2002d:4).

that these calls were due to the criticism of both the SPD and the PDS when they formed their first coalition government in 1998. Still, we find the opposition party as well as sub-regional actors maintaining that joint action, unity and 'the common good' should come to the fore in regional developmental work.

If starting our exposition at the subregional level, we will learn that the unification of political and economic forces is considered a central element in the work on developing the economy in the subregions. In the Regional Development Agenda (RDA) for Mid-Mecklenburg/Rostock, it is said that the joint search for 'regional solutions and endogenous potentials' constitutes the most important part of the work on regional development. Joint efforts, it is maintained, create unanimity, stimulate the mobilization of hidden potentials in the region, and enhance the attractiveness of the region in more general terms. Additionally, it is argued that 'only those who conceive themselves as a region and are ready to tackle the challenges with joint forces' will be able to succeed in the global competition between regions.[54] In the RDA for Mecklenburg Lake District, *Kirchturmpolitik*[55] and the absence of a common focus are cited as weaknesses of this particular region.[56] It should be observed that the *Land* government does have a say also at the subregional level. The RDAs, for instance, should not conflict with the initiatives taken or the regulation provided at the level of the *Land*. In yet other contexts than the RDAs, however, it is claimed that solving regional problems will be possible only through cooperation and broad regional consent.[57]

Alongside these subregional voices, CDU representatives bring out the need for joint effort and visions.[58] In a publication from 1996, the CDU refers to the time of *Die Wende*, when the social and political climate was 'marked by a strong polarization'. Now it is time to reconcile the conflicts and the 'mental walls' between East and West and between different parties in the region.[59] In a publication from 2002, the CDU stresses that politics need to overcome ideological barriers and to start focusing on the long-term interest of the *Land*.[60]

---

54 RPV Mittleres Mecklenburg/Rostock (2002:3,6).

55 'Parochialism'.

56 RPV Mecklenburgische Seenplatte (2000a:43).

57 Daedelow, Dietrich (2000), in RPV Meckleburgische Seenplatte (eds.), *REGIOnal*, No1, Neubrandenburg, p.9; Kautz, M. (2000), in RPV Meckleburgische Seenplatte (eds.), *REGIOnal*, No 3, Neubrandenburg, p.1.

58 CDU MV (2001)

59 CDU Landtagsfraktion MV (1996:5).

60 CDU Landtagsfraktion MV (2002:8).

There is, it is argued, a need for a common and positive *Zukunftsbild*,[61] which should be capable of mobilizing all public and social interests in the *Land* and of attracting investors to the *Land*.[62] During the election campaign from the same year, the emphasis on joint efforts is energetically maintained. In one election manifesto, the party launches the idea of forming an 'alliance for Mecklenburg–Western Pomerania': 'it is only by acting together that we can get the *Land* going'.[63] In one of his speeches, the CDU party chairman invites 'everyone who want to participate wholeheartedly in the efforts with getting the *Land* going' to work on developing the *Land*; it is 'engagement with the *Land* rather than the party label' that needs to come to the forefront in political work, and the *Land* will develop only if politicians start working 'in a spirit of partnership'.[64] At the end of the same speech, the assembly is exhorted to use the few months before the elections 'to fight and to argue – resolute and jointly – for the sake of the *Land*!' The final point made in the speech certainly summarizes much of its message: '*Für unser Land!*'.[65]

Growth and Unity as Core Values

To judge from these examples, it is neither political autonomy, cultural recognition nor more extensive state subsidies that constitute the overriding goal in the two regions studied. As an alternative, it is regional economic growth and intra-regional unity that stand out as core values in Norrbotten as well as in Mecklenburg–Western Pomerania. Competitiveness and endogenous growth are referred to and represented as interests of the entire regional population. Just like unity and common stances, the other core values communicated, these features are promoted vital yet essentially unproblematic values.

The Changing Nature of Values

Before considering the implications of adopting 'growth' and 'unity' as core values, we need to discuss whether it is at all possible to classify a policy that lacks the claims for increased autonomy as an instance of regionalism. Hence,

61 'Vision of the future'.
62 CDU Landtagsfraktion MV (2002:5,8).
63 CDU MV (2002:2).
64 Rehberg (2002).
65 'For our Land!', Rehberg (2002).

the claim for some kind of autonomy has in many cases been conceived as a defining element of regionalism.[66] Here, though, the absence of claims for self-governance or autonomy must be understood in context. There is nothing to show that the core values of regionalism always have to be the same. To the contrary, it is likely that the core values defended will change concurrently with the context. The absence of claims for self-governance observed here may hence due to the so-called changing significance of territory.[67] Thus, today, variations in economic development and growth are, as indicated in Chapter 4, often explained by reference to the regions' soft location factors, to their 'endogenous potentials' and to their specific assets. This implies, as Keating points out, that the key powers in regional development are no longer thought to be those exercised by the central state. The focus on know-how and learning increase, as an alternative, the importance of policy areas such as education and training, of which many are already managed at the regional level.[68] These circumstances, as well as the belief that the state is on decline in quite general terms, have contributed to making political autonomy less important. Instead, endogenous growth and intra-regional unity have joined the regionalist core values.

Competitiveness and Growth as Political Objectives

That regional representatives are preoccupied with trying to foster economic growth does not come as a surprise. Economic growth is thought to solve profound social problems, particularly in less favoured regions. The 'better society' sought for in the regions could thus be one in which the region has a strong economy, a high level of foreign investment, low unemployment rates and positive migration flows. In this 'better society' growth would be a means to an end rather than an end in it self. Yet the two regions studied here do illustrate a so-called 'reconfiguration of politics ... around the theme of growth and regional competitiveness'.[69] To 'catch up with the national average' has then become a nodal point around which much policy-making in the two regions revolves.[70] Growth and development are no longer represented only as instruments for reaching some certain ends, but have evolved into higher values that have

---

66 See Chapter 1.
67 Keating (2004:373).
68 Keating (2004:374)
69 Keating (1997:29).
70 Halkier, Henrik (2004), 'Institutions, Discourse and Regional Policy', *Lecture at ECPR Summer School*, Aalborg.

achieved a central position in the sense that other values are defined in relation to them.

To adopt 'competitiveness', 'development' or 'growth' as key political objectives is certainly not without problems. For example, these features are both elusive and hard to measure in an unambiguous way.[71] Different measures of growth and development may, as Keating suggests, simply 'conflict with each other'.[72] Employment and productivity are just two examples of measures of development that are not necessarily compatible with each other. Another problem with adopting 'competitiveness', 'development' or 'growth' as political core values is, as argued in Chapter 4, that it is *companies* that are or are not competitive, not the regions they happen to be located in.[73] Companies located in the same region can actually compete first and foremost with each other. Talking about competitive *regions* is therefore quite misleading.

It is probably due to, and not in spite of, the inherent vagueness of the concepts that 'competitiveness', 'development' or 'growth' have been accorded such a central position in the regional policy discourse. Thus, regional representatives can talk about development and growth in rather general, sweeping or imprecise ways, and without knowing whether their understanding of them is shared by others. The concepts are hardly ever thoroughly discussed or analysed. Instead, they are used as catchwords around which many actors in the region can unite. The problems with adopting 'growth' and 'competitiveness' as goals for a territorial or political unit are, however, widely overlooked in the political debate.

Unity and Common Interest as Political Objectives

The political debate analysed here does not reflect any particular conflict of interest between the state and the regions. We saw indeed a handful of examples of discontent against the state being voiced in Norrbotten. It is not yet a regionalism fraught with conflict that we encounter here. Instead, one of its defining elements is the ambition to form a community of interests *within* the region. The tendency among regional agencies to strive to 'promote and defend the interest of "their" areas in the face of increasing competition for investments and re-

---

71 Keating (1997:31).

72 Keating (1997:31). See Mankiw, Gregory N. (2004) *Principles of Economics*, Thomson South-Western, p.499-517 for a comprehensive discussion about ways to measure economic growth.

73 Lovering (1999:388-9).

sources'[74] is palpable in both Norrbotten and Mecklenburg–Western Pomerania. Calls for unity and concord are found among conservatives, social democrats and socialists alike. in these regions as well as in many other regions in both Sweden and Germany. Several efforts are made, particularly in Norrbotten, to reach a regional consensus. Activities such as seminars and meetings are arranged in order to find common stances on central issues. There is, in fact, a tendency in the regions to unite around the importance of being united. This, I would say, could thus be understood as 'double layers of unity' in the regional political debate.

Even if some of the calls for unity just end up in a consensus on the value of unity, there seems to be faith in the possibility of pursuing a policy that is for the good of the entire regional population. In both regions, we come across examples of how previous or ongoing conflicts and disagreements are played down and replaced with a call for unity. Pragmatic solutions that overcome ideological and party political boundaries are conceived as an ideal, and politicians of a less disputatious kind are asked for. Ideological differences are regarded mostly as problematic hindrances to growth and development, whereas unity and mutual understanding are thought to foster economic growth.

This policy of mutual understanding does give rise to several reflections. Initially, it could be disputed whether this policy is 'new' in any respect or just a continuation of existing policy. Hence, it has been argued that compromise politics and consensus democracy have been particular features of Swedish political life throughout the twentieth century. According to Lewin, however, earlier forms of consensus politics were still based upon a latent conflict that could be activated by the Social Democratic government and transmitted to the non-Socialist parties and to the trade and industry whenever it was needed. Now, it is argued, even latent conflict is absent. Economic liberalism dominates the political agenda, welfare politics is increasingly called into question, and 'the non-Socialists have', as Lewin puts it, 'taken over much of the ideological initiative'.[75]

This policy of mutual understanding, furthermore, throws new light on the classical political question of 'who gets what, when and how.'[76]. Political scientists have regarded this question as one of the core problem of politics, and in a situation of scarce resources these matters are of course brought to a head, as limited resources imply that some groups and goals must be given priority over

74 MacKinnon *et al.* (2002:296).

75 Lewin, Leif (2002) *Bråka inte!*, Stockholm: SNS Förlag, p.27.

76 Hague *et al.* (1992:3).

others.[77] In Norrbotten and Mecklenburg–Western Pomerania, however, the question of conflicting interests at the allocation of resources does not seem to play the most central part in the public political debate. The main focus is not on how to prioritize between clashing interests, but on how to form an interest that is common for all. And the question is no longer how social and public resources should be allocated, but how they should be created. This links up to an analysis made by Harvey while studying metropolitan cities. Urban governance, it is argued, has become 'increasingly preoccupied with the exploration of new ways in which to foster and encourage local development and employment growth'. Thus, the 1960s so-called managerial approach to urban governance was already in the 1980s increasingly outmanoeuvred by an 'entrepreneurial' form of action.[78] In the light of that, it seems as if one of the core problems of politics – the authoritative allocation of values[79] – in many cases has had to make way for a policy aimed primarily at *creating* value.

The Region as a Community of Common Interests

Among all the values that can be found within the regions, there is an established order that makes some values more central than others. In the regions studied here, unity is represented as a high-ranking value in the regional policy discourse. Yet unity is also clearly linked to, and defined in relation to, endogenous growth and competitiveness, which is 'top of the value charts' in both regions. To group the whole region together, and to imply that the whole regional population would benefit from the same policy, is, of course, highly problematic. Initially, it could be objected that the talking about *regional* 'competitiveness', 'development' or 'growth' largely neglects the fact that these features are never distributed evenly over regional space or among the regional population.[80] Parts of a region can experience an economic boom, whereas other parts lag behind.[81] And a change in the economy may benefit one section of the regional population while it harms other sections. Most development strategies are 'likely to produce

---

77 Hague *et al.* (1992:3f.); Held, David (1996), *Models of Democracy*, Cambridge: Polity Press, p.211.

78 Harvey (1989:3).

79 Lundquist, Lennart (1993), *Det vetenskapliga studiet av politik*, Lund: Studentlitteratur, p.27-28.

80 Lovering (1999:384).

81 Lovering (1999:388-9).

winners and losers within the region'.[82] Strategies that, like those found in Norrbotten and MV, focus on competitiveness and growth have a tendency to 'increase social inequality, since resources are diverted away from social programmes to those activities which promote growth'.[83] By defining growth and competitiveness as a 'common regional interest' it is, as Keating points out, possible to narrow the political debate and to justify policies that are inherently pro-business.[84]

The mere idea that there is a common regional interest of any kind can, furthermore, be countered by the objection that geo-political units such as Norrbotten and Mecklenburg–Western Pomerania are constituted by a multitude of diverging groups and interests. To suggest that all of their inhabitants would benefit from a certain policy is, therefore, highly controversial. The regionalist idea about a common regional interest is also suggestive of the nationalist idea about the *nation's* best interest. Hence, nationalism can be described as an ideology, a political movement, or even as a form of behaviour which seeks to defend and promote the interests of the nation.[85]

It can indeed be objected that most political and economic interests in society have tended to legitimize their existence by claiming to represent the 'national interest'.[86] Populists as well as representatives of established political parties have been noted to 'play the patriotic card' to claim that 'their politics, and their alone, has the nation's interest at heart'.[87] Here, we have observed how regional representatives from all parties claim to pursue a policy that is for the *region's* good. This may be understood as an attempt to use the 'regionalist card' merely for strategic ends. If that was the case, however, the politics that was to be 'sold' or promoted by means of the 'regionalist card' would take different shapes, depending on the political preferences of the 'card player'. Here, this does not seem to be the case. There seems, rather, to be a real consensus around the value of both unity and endogenous growth. The 'regional interest' is thus neither a phrase used for selling any political idea, nor an 'abstractly formulated moral

---

82 Keating, (1998:143).

83 Keating (1997:30).

84 Keating (1998:143).

85 Kellas (1998:4); Gellner, Ernest (1997), *Stat, Nation, Nationalism*, Nora: Nya Doxa, p.11.

86 Löfgren, O. (1993a), 'Nationella Arenor', in Ehn, B., Frykman, J. and Löfgren, O (eds.) *Försvenskningen av Sverige*, Stockholm: Natur och Kultur Förlag, p.148f.; White, Richard (1981), *Inventing Australia: Images and Identity 1688-1980*. Sydney: Allen & Unwin.

87 Billig (1995:103).

principle' with no repercussions on political life in the regions.[88] Instead, the idea of 'the region's best interest' provides its advocates with 'a guide for action'.[89] Thus, even those who represent parties that are critical of capitalist competition or repudiate the rush for economic growth suddenly become advocates 'for the very process they are trying to resist'.[90] These actors perform, however, not only as actors for 'capitalist competition' but as region-builders who focus on the imagined interest of a population that in reality is quite heterogeneous in their interests and demands.

It is true, however, that attempts to speak for the whole region won't work unless there is a common and familiar idea about what the region is or who 'we' are in a particular context.[91] Any attempt to speak for the entire region must therefore fall flat unless there is a widespread idea about what the region is. There are, however, many attempts to bring about an *image* of the region as a coherent space. These attempts are further discussed in Chapter 7.

88 Pin-Fat, Véronique (2005), 'The Metaphysics of the National Interest and the "Mysticism" of the Nation-State: Reading Hans J. Morgenthau', *Review of International Studies,* 31, pp.217–236.

89 Pin-Fat (2005).

90 Harvey (1989:5).

91 Billig (1995:103).

6. Organizational ideals of Regionalism

After establishing that many political actors in the two regions describe the world as globalized, regionalized and competitive, we learned that economic growth and intra-regional unity constitute two main core values in the regional policy discourse. However, regionalism is, like other ideologies, also action-oriented and tends explicitly to link values to action.[1] Regionalism involves, as suggested above, an attempt to manipulate the political process allegedly in favour of the region and its inhabitants.[2] But this manipulation of the policy process can be inspired by totally different organizational and strategic ideals.[3] Regionalist and nationalist ideologies have often adopted radical or even violent political ideals. Northern Ireland, the Basque Country and Corsica have been mentioned in a number of contexts as examples of places in western Europe where political goals with sub-nationalist or regionalist overtones have been promoted by violent means. Yet new regionalism has been characterized by an inclination to use diplomacy and peaceful lobbying. New regionalism first emerged in regions that have been known for possessing the political power and the financial means to lobby in the international arena. We therefore need to inquire into the political methods advocated in our two regions. By what means is the political process said to be manipulated, and how are these means represented and justified in the regional policy discourse? What organizational and strategic ideals are communicated to people in the two regions?

NORRBOTTEN

Globalization, Europeanization and an alleged dismantling of the central state are, as indicated in previous chapters, thought to confront the region with new challenges. In Norrbottnic political life, it is frequently suggested that these challenges are best met by cooperation among different partners in the region.

1 Ball and Dagger (1999:7); Rejai (1995:9); Funderburk and Thobaben (1989:4).

2 Williams (1997:114).

3 Hansen and Saglie (2005); van Dijk (1998:70); Skovdahl (1992:43); Funderburk and Thobaben (1989:4); Baradat (1991:9).

This argument came to the fore in *The Guide*, in the debate about a proposed restructuring of the regional institutional order,[4] and in a multitude of other contexts.[5] Political texts from Norrbotten contain calls for joint action and alliances between 'different actors', between and within municipalities, between private and public actors, between small and large companies and between different lines of business.[6] In many of these cases, it is suggested that the cooperation should take the form of a network or a partnership. Networking and voluntary cooperation are in several cases said to be essential for the competitiveness and development of the region.[7] In a Norrbottnic context, several partnerships and networks could be cited as examples of how the desire to join forces has manifested itself. In the following sequence, some of them will be brought out and discussed. The first example is the *Norrbotten Group*[8], the second is *Europaforum Northern Sweden* and the third is *Norrbottnic International Council* (NIC). The discussion below does not concern the ways in which these partnerships really have worked, but how they have been justified and represented in the regional policy discourse.

The Norrbotten Group

Arguably, the European Union once introduced the partnership principle at the regional level in Sweden. In Norrbotten, however, the tradition of cooperative partnerships is older than Sweden's membership of the EU. During the summer

4 This debate has been interesting to follow, even if no formal change of the Norrbottnic decision-making bodies was accomplished during the period to which this study is confined. The debate was very much influenced by the so-called regional experiment in southern Sweden, where a handful of pilot regions were both restructured and given extended responsibility for regional development policy. See Jerneck and Gidlund (2001). See also Chapters 3 and Chapter 5, this volume.

5 Regionutredningen Norrbotten (2004), *Handlingskraft - Delaktighet och Kraftsamling för Norrbotten*, Luleå, p.3ff.

6 Folkpartiet Liberalerna Norrbotten (2002a:5); SAP Norrbotten (1997:4); Vänsterpartiet Norrbotten (1998:4); SAP Norrbotten (2001:3); Ström (2002); Forsberg, Robert (2003), 'Samverkan är enda vägen', *NSD*, 2003-04-30; Regionutredningen Norrbotten (1999b:8), Kommunförbundet Norrbotten (2000:3,6); NLL (1996a:14); NLL (1997c:17-18); NLL (1998c:23,28); NLL (2000b:14); Öberg (2002); Europaforum (2003); NSD (2001b), 'De vill få slut på revirtänkandet', 2001-09-26.

7 Öberg (2002); Eriksson (2003b); Regionutredningen Norrbotten (1999a); Kommunförbundet Norrbotten (2000); Europaforum 2003). NLL (1999:14).

8 *Norrbottengruppen* in the original.

of 1993, the County Administrative Board, the County Council and the Regional Association of Local Authorities (RALA) in Norrbotten entered into a dialogue concerning a common project aimed primarily at discussing the future development of the county. Initially, the cooperation between the parties went under the name 'Partnership for Norrbotten'. Since 1996, however, the group has been called 'the Norrbotten Group'. The three actors initiated the partnership, but aimed to include other parties in the work, too. The Norrbotten Group engaged a number of experts and gradually two consultative groups were connected to it – one made up of representatives from the municipalities and the other of representatives of trade and industry and the trade unions.[9]

One of the first assignments of the group was to draw up a strategy for economic development in the region. This work went under the name *Strategi 95*, and has been described as a turning point in regional developmental work in Norrbotten. A wide range of actors were involved in preparatory work, that is, in analysing the preconditions for growth and development in the region. The result of that work was contained in a number of reports, later presented at seminars to which the participating organizations, the media, and the general public were invited. The final result of this work was *The Guide*, already much quoted in previous chapters.[10] When regional actors refer to *The Guide*, they often stress that the dialogue and the cooperation that emerged in the preparatory work was even more important that the final publication. In an evaluation of *The Guide* commissioned by the Norrbotten Group, the document is said to amount to the beginning of a 'new way of working with regional development issues', having 'cooperation as the guiding spirit'.[11] When the County Council reports on the role *The Guide* has played in the development work in the region, it too emphasizes the process of cooperation and joint analysis.[12]

The Norrbotten Group turned out to be of importance even after its work on *The Guide* was complete. It played a central part in the formulation of a Regional Growth Agreement[13] (RGA) for Norrbotten, and took part in drawing up the

---

9 Regionutredningen Norrbotten (1999a:5); NLL (1996a:11).

10 For a more detailed account for the making of *The Guide*, see Chapter 4.

11 Björnemark, Maria (1998), *Vägvisaren, hur har den uppfattats av Norrbottens aktörer?* Luleå: CUFS.

12 NLL (2002b), 'Vägvisaren, en välkänd strategi', *Pressinformation*, Luleå.

13 Regional Growth Agreements (RGAs) were introduced by the Swedish government in order to stimulate better collaboration between actors at local, regional and central levels. The RGAs were also presumed to increase local and regional influence over state policies at the regional level. The RGAs were drawn up during 1999 and became operative the year after, after being approved by the Swedish government.

Single Program Documents (SPD) for northern Sweden.[14] The cooperative and partnership working methods used in preparing these documents have been considered a major advantage. In the 2000–2 RGA, it is argued that the preparatory work in producing the agreement was an important process in itself. The 'gathering around a joint analysis and a common developmental aim and direction' is said to have provided a 'basis for an energetic policy'.[15] In the subsequent Regional Growth Programme (RGP), the ambition to 'influence attitudes, and to form a common value-system as regards the direction and the form of the regional developmental work' is brought out as being of the outmost importance.[16] According to the County Administrative Board, work on the agreements has resulted in a 'genuine revitalization' in the region, in a 'common focus', in win-win situations, participation, cooperation, engagement, trust and optimism.[17] The SPD for northern Sweden focuses, like the RGP, on cooperation, networking and joint action. Cooperation – between trade and industry, research and development, municipalities and local actors – is understood as one of the most important potentialities for northern Sweden.[18] In the SPD as well as in the RGP, it is said that teamwork and communication between the business community and the surrounding society should be further developed.[19]

Partnerships for International Affairs

A common assumption among policy-makers in Norrbotten is that internationalization requires cooperation across borders. In the 1998 International Action Program (IAP) for Norrbotten, internationalization is said to demand new coalitions and new forms of cooperation among regions, companies, authorities and individuals.[20] Cooperation across borders, whether it is with other regions or with different bodies within the EU, is motivated by the argument that problems shared with others are best solved in cooperation with

---

14 See NLL (1998c:7).

15 Norrbottengruppen (1999:5f.).

16 Norrbottengruppen (2003:38); Lst Norrbotten *et al.* (2000), *Samlat Program Dokument, Mål 1 Norra Norrland*, 2000-05-24, Luleå and Umeå, p.43.

17 Lst Norrbotten (2001a), 'Länsrallyt i Norrbotten' at *www.bd.lst.se/news/tvavtal/010606.htm* Luleå, 2001-06-06; See also Lst Norrbotten (2003c:37) .

18 Lst Norrbotten *et al.* (2000:32ff).

19 Lst Norrbotten *et al.* (2000:43); Norrbottengruppen (2003:19).

20 Norrbottengruppen (1998:6).

them.[21] Cooperation, it is argued in the IAP, takes place not only because regions 'need each others' support to handle common challenges', but also because of 'cultural and other similarities'.[22]

Internationalization is, moreover, thought to require a higher degree of unity *within* the region itself. We find regional politicians arguing that they must learn to 'speak with one voice' in international communication, that it is important to find 'common Norrbottnic positions on different international questions', and that it is vital to give the region a profile in the international arena. The importance of 'being heard' at the European level or of 'being present where decisions are made' is pointed out in several contexts.[23] A Norrbottnic County Governor stresses that 'northern Sweden must show up more often in Brussels' and that 'local and regional politicians and representatives must go to Brussels to participate in the debate about the EU's future regional policy'.[24] European integration in general and the changing distribution of the union's structural funding in particular are emphasized as reasons why representatives for the County must get together and become more internationally active.[25] Cooperation and partnership thus appear to constitute an organizational ideal also when it comes to creating and keeping international contacts.

The idea that 'one must be present where decisions are made' and the belief that international engagement is decisive for the region's future development have contributed to many regions' decisions to establish permanent representation in Brussels.[26] The County of Norrbotten is no exception. Since September 1997, a wide range of public and private Norrbottnic partners have run the North Sweden European Office (NSEO), together with the adjacent

---

21 Rusiniemi, Anna-Mari (2002), *Den nya regionalismen i Europa – en studie av Norrbottens och Västerbottens ställningstagande till den Europeiska Unionens sammanhållningspolitik*, Bryssel: NSEO; Hedkvist Petersen, Eva (2002), 'EU-geografin kan inte harmoniseras', *NSD*, 2002-06-08; Lst Norrbotten *et al.* (2000); Rosengren (1998).

22 Norrbottengruppen (1998:5).

23 NSD (2002b), 'Företagen får *en* röst', 2002-12-21; NIR (2003a), 'Minnesanteckningar' at *nir.norrbotten.se/prot/prot_2003-03-20.htm*, 2003-05-16; Norrbottengruppen (1998) *Internationell Strategi för Norrbottens län*, Luleå; Lst Norrbotten (2001b) *Internationellt Handlingsprogram*, Luleå

24 Marklund (2003).

25 Westman (2002); Lavander, Per and Öberg, Maria (2002), 'Nya utmaningar i Europas utkant', *NSD*, 2002-03-05.

26 Burch and Gomez (2002); Gren (2002b).

County of Västerbotten.[27] NSEO is, like most regional representation offices, supposed to monitor policy changes considered to be of consequence for the regions of Norrbotten and Västerbotten, and to communicate these changes to what is called 'the relevant forum in the regions'.[28] The organization is also said to be responsible for forging alliances with other regions and to form networks with regions whose interests and problems are similar to those of northern Sweden.[29] The office in Brussels has also been instructed to 'put Norrbotten and Västerbotten on the map' and to introduce the region to 'important agents in Brussels and the rest of Europe'.[30]

One of NSEO's priority tasks – to supply decision makers in Brussels with perspectives 'from Northern Sweden' – has, however, not been without problems. There is no common political representation of the northernmost counties in Sweden, and consequently it was not always clear whose point of view NSEO should communicate to Brussels. In the meantime, similar problems became evident at Mid-Sweden Office, the permanent representation for the other two northern Swedish counties. This prompted the establishment of *Europaforum Northern Sweden,* a platform where common policies for the four northernmost counties in Sweden could be formulated.

Europaforum Northern Sweden

Europaforum was founded in 1999 and gathers political representatives from the regional and local levels of Sweden's four northernmost counties. The forum has then been described as a 'partnership' or as a 'mutual political mustering of strength'.[31] The purpose of the partnership is said to be to strengthen the position of the northernmost counties in the EU and to define common stances on the

27 The founding organizations are the Municipalities, the County Councils, the County Administrative Boards, the Federation of Private Enterprises and the Chambers of Commerce in Norrbotten and Västerbotten, as well as Norrlandsfonden, Luleå University of Technology and Umeå University. See NSEO (2003a), 'About us' at *www.northsweden.org/english/aboutns/aboutns.htm*, 2003-08-19.

28 NSEO (2003a).

29 NSEO (2001), *Verksamhetsplan för år 2002*, Bryssel, Luleå, Umeå.

30 NSEO (1999), *Verksamhetsplan för år 1999-2000*, Bryssel, Luleå, Umeå.

31 Europaforum Norra Sverige (2001) *Positionspapper om Europeiska Unionens framtida sammanhållningspolitik*, Lycksele.; Europaforum Norra Sverige (2003a), *Deklaration från Europaforum Norra Sverige V*, Östersund; Europaforum Norra Sverige (2003b), 'Gemensam politisk kraftsamling', at *www.europaforum.nu*, 2003-09-26.

EU's future cohesion policy.[32] It is also to 'develop the teamwork between politicians at local and regional levels in the four, northernmost counties in Sweden' and to exert an influence on the parts of EU policy that concern this area.[33] There is, according to Europaforum, 'a great need for cooperation and joint action' in making contact with national and European authorities, and with international organizations and other regions in Europe.[34]

In principle, the meetings of Europaforum are open to everyone interested, and as a result we find private actors as well as a considerable number of civil servants taking part in the meetings. Still, the main aim of the forum is to achieve political anchorage for lobbying efforts in Brussels and, as a consequence of this, the gatherings are arranged primarily for politicians at local and regional levels. Since the working principle of the forum is voluntarism, the composition of its political majority can differ from meeting to meeting. The forum lacks formal authority, which prevents it from making binding majority decisions. Instead, the forum makes so-called open decisions arrived at by consensus: that is, as long as no participant dissents, a proposal is understood to enjoy undisputed support. The main responsibility for securing the counties' and the municipalities' support for the decisions made at the forums is left to the delegates, as is the responsibility for getting the decisions implemented.[35] One of the initiators of Europaforum stressed in a speech that the procedure at Europaforum's meetings 'is not to vote, but to talk things over'.[36]

Europaforum meets twice per year, but more intense work is done by the rapporteurs – a minor group of delegates working together to prepare the meetings and to execute the commissions given to them. The rapporteurs are, moreover, supposed to gather information on 'significant topics', and to forward this information to the 'relevant actors' in their respective counties.[37] The rapporteurs are designated to represent different parties and different counties. In discussions regarding their role, however, it is argued that their party label or their geographical origin should not matter for the positions taken by them. Since the assignment of the rapporteurs is to strive for consensus, this kind of

32 Nilsson, Jens (2002a), 'Inledningsanförande', *EuropaForum IV i Sundsvall*, 2002-06-05.

33 Europaforum Norra Sverige (2002a), *Inbjudan till Europaforum Norra Sverige IV*, Härnösand, p.2.

34 Europaforum Norra Sverige (2002b), *Förslag till fortsatt inriktning*, Sundsvall, p. 1.

35 Nilsson, Jens (2002a).

36 Nilsson, Jens (2002b), 'Anförande', *EuropaForum IV i Sundsvall*, 2002-06-05.

37 Löfstedt, Marianne (2002), 'Rapport från rapportörsgruppen', *EuropaForum IV i Sundsvall*, 2002-06-05.

convergence of political thought is not seen as a problem. It is, to the contrary, argued that the ideal is that it should be impossible to judge from their arguments from which party or from which county the rapporteurs come.

The cooperation and the unity between the participating actors are described not only as 'valuable' but also as 'awfully fragile'.[38] The ability to agree on minor issues is put forward as important if the forum is going to 'agree on big issues in the future'.[39] The partnership model, the working with *rapporteurs* and the striving at consensus are – the fragility notwithstanding – still considered as a possible model for other policy-making institutions in northern Sweden.[40] The Regional Association for Local Authorities and the County Council contend that having rapporteurs 'working for broad support and consensus' is a convenient organizational principle.[41]

The Norrbottnic International Council

Besides Europaforum, we find additional organizations set up to achieve intra-regional unity in international affairs. One example of such an organization is the *Norrbottnic International Council* (NIC). NIC was founded by the County Council and the municipalities in Norrbotten at the end of 2002, and is supposed to function as a network for internationally active politicians from different parties within Norrbotten.[42] The network is thus not open to anyone interested, but is accessible primarily by politicians holding a mandate in any of the international organs in which the county is represented. At the beginning of 2003, NIC was made up of a dozen politicians and a good handful of civic servants, aiming to meet a few times each year.

The role of NIC is described as to 'exchange information' and to 'spread knowledge about how important international activity is for regional development'. It is maintained that the NIC should *not* be a 'lobbying organization' for Norrbotten.[43] Nonetheless, it is maintained that the network aims at 'laying the ground for a common Norrbottnic position on different

---

38 Nilsson, Jens (2002b).

39 Andersson, Inge (2002), 'Anförande', *EuropaForum IV i Sundsvall*, 2002-06-05.

40 Nilsson, Christer (2002), 'Anförande', *Europaforum IV i Sundsvall*, 2002-06-05.

41 Regionutredningen Norrbotten (2002), *Region Norrbotten – samverkansorgan för Norrbottens län*, Luleå, p.18; NLL (2002a), 'Yttrande över EU-kommissionens vitbok om styrelseformer i EU', *Dnr 677-02*, Luleå.

42 NLL (2002c), 'Norrbottens Internationella Råd bildat', *Pressinbjudan*, Luleå.

43 NIR (2003a), *Minnesanteckningar 2003-03-20*, Luleå.

international questions' and that Norrbottnic politicians must learn how to 'speak with one voice'.[44] According to the NIC, internationalization increases the importance both of regions and of popularly elected regional politicians. The 'international perspective' is described as being 'as important and natural as the local, regional and national perspective', as a precondition for regional development, and as essential for turning Norrbotten into an 'attractive and competitive region'.[45]

Broad support for Cooperation

The emphasis on cooperation, partnerships and networks is a common feature in regional developmental policy, and criticism of this way of working appears to be minuscule. Apart from formalized partnerships and networks, many other initiatives are taken to bring regional actors together.[46] We find initiatives such as 'Days for Dialogue' and 'Regional Development Days', thought to stimulate dialogue between the business sector and politicians in the region, to create a 'positive atmosphere in the region' or to function as a means of finding 'common visions' or a 'common value system' for regional developmental work.[47] A Norrbottnic County Governor contends that 'the ability to cooperate and form networks' is characteristic of 'dynamic' regions such as northern Italy, and Gnosjö and Karlskrona in southern Sweden. One important factor behind the success of these regions, it is argued, is the fact that 'a variety of actors take part in a joint discussion about the development of the region'.[48] Also, a succeeding County Governor maintains that 'the capacity for cooperation is what unites most sucessful growth regions'.[49]

Cooperation is recommended as the best practice by the opposition as well as by the majority in the County Council. In its 2002 election manifesto, the Liberal Party stresses that regional developmental policy must focus on decentralization of the state and on local cooperation.[50] A Social Democratic representative holds

44 NLL (2002c); NSD (2002b), *'Företagen får en röst'*, 2002-12-21; NIR (2003d), 'Ett nätverk för politiker', at *nir.norrbotten.se*, 2003-10-27.

45 NIR (2003c), 'Rådets Uppgift' *at nir.norrbotten.se/syfte.htm*, 2003-10-28.

46 NLL (1997b), *Dialog för ett bättre företagsklimat*, Dnr Lka1250-97, Luleå.

47 NLL (1997b); Marklund (2000); Lst Norrbotten (2001a).

48 Marklund, Kari (2002), *'Samspelet avgör den regionala utvecklingen'*, NSD, 2002-10-31.

49 NIR (2003b).

50 Folkpartiet Liberalerna Norrbotten (2002a).

that 'cooperation, cooperation and cooperation again' is the main ingredient of regional success.[51] A representative for the Left Party contends that cooperation and joint action between municipalities, trade and industry, public administration, political forces, associations and the educational system is a precondition for regional success.[52] Another representative for the Left Party is reported to have said that institutionalized partnerships and networks should be complemented with informal and spontaneous ones.[53] A representative for the Centre Party, however, criticizes the tendency to work in networks and partnerships for leading to a dominance of older men in politics. Still, it is argued, new networks host 'unconventional ideas and a vitalizing power', from which the established networks could learn a lot.[54] A representative for the Moderate Party stresses, furthermore, that the departmentalization of responsibilities needs to be replaced by cooperation and joint action. In fact, it is argued, this ought to be 'a matter of course for organizations' as well as for enterprises and team games.[55]

Most actors, as indicated above, tend to understand cooperation between private, public and non-profit sectors as a precondition for regional economic development. The involvement of the private sector in regional developmental work is defended with the argument that this work aims at creating propitious opportunities for companies, and that private entrepreneurs are best suited for deciding what efforts are the most propitious for them.[56]

MECKLENBURG–WESTERN POMERANIA

The government in Mecklenburg–Western Pomerania is, as pointed out in previous chapters, rather explicit in its search for joint visions, strategies and efforts. In particular, the PDS Minister for Regional Development makes a strong a case for 'strategic alliances', 'gigantic joint efforts' and other forms of cooperation. The alliances referred to are in some cases said to include partners from the social sector as well as from trade and industry, scientists, artists, publishers, trade unions, Chambers of Commerce, churches and municipalities.[57]

---

51 Ström (2002).

52 Lööw, Börje (2002), 'Skippa skitsnacket och tala om utveckling', *NSD*, 2002-07-11.

53 Regionutredningen Norrbotten (2000a), *Seminarierapport 2000-01-27*, Luleå, p.1.

54 Regionutredningen Norrbotten (2000a:1).

55 Hultin, B. (2004) 'Samverkan och positivt tänkande ger tillväxt', *NSD*, 2004-09-10.

56 Regionutredningen Norrbotten (1999a); Lst Norrbotten *et al.* (2000).

57 MfABL (2002c).

Here, *Denkwerkstatt 2020* (DW 2020), established by the PDS Minister of Regional Development in April 2000, will be discussed as one example of how this idea has been put into practice in MV. The bid for partnerships and development coalitions is strongly evident also at the subregional level. In the four subregions in Mecklenburg–Western Pomerania, we find several examples of how the organizational ideal of cooperation and joint efforts has been implemented and argued for. In three out of four planning regions, networks, or partnerships aiming at *promoting* the subregion, have been established. Another and even more illustrative example of subregional cooperation is, however, constituted by the so-called Regional Management Initiatives. Below, the Regional Management Initiatives as well as the networks aiming at promoting the subregions will be further discussed.

Denkwerkstatt 2020

Denkwerkstatt 2020 (DW 2020) could be mentioned as one example of how the organizational ideal of broad alliances and cooperation has been institutionalized in political life in MV. DW 2020 is described as a 'think tank' to which 'prominent guests' are invited in order to 'gain experience, discuss and to find possible paths to development'.[58] When the think tank was established, the Minister for Regional Development stressed that there is a need for a 'long-lasting and reflexive style of politics'. Politics, it was argued, should have a perspective that 'goes beyond legislative periods and aims at the best for the *Land*'. DW 2020, it was further contended, should serve these purposes and at its half-yearly meetings come up with new visions, ideas and strategies for the *Land*.[59] The *Land* was thus said to benefit from a joint analysis by a wide range of individuals representing different spheres in society. It is, according to the Minister, only through broad cooperation that 'new and joint ideas, capable of solving the problems of the *Land*' can emerge. It is stressed that the analytical work done within the think tank must adopt 'a pragmatic approach', and that the visions worked out need to have strong support also outside traditional political parties.[60] The fact that the participants (that is, representatives of banking, trade and industry, the university and political parties) have declared themselves

58 MfABL MV (2001b), *Denkwerkstatt 2020 – Chancen für Mecklenburg-Vorpommern*, Schwerin, p.2.

59 MfABL MV (2000a).

60 MfABL MV (2001b:2,24).

willing to take part in this 'cross-party' project should, according to the responsible Minister, lead to several useful 'pragmatic recommendations' for the decision-makers in the *Land*.[61] It is also stressed that a successful regional developmental policy needs to be worked out in cooperation with scholars and with support from the business sector. Sustainable visions of the future can, according to the Minister, emerge only in cooperation across the traditional borders of policy-making.[62]

Regional Management Initiatives

Since the beginning of January 2001, several *Länder* in Germany have taken part in the federal Regional Management Initiative. The project was initiated by the Federal Ministry of Economics and Labour and has been subsidized by the *Land* as well as by the federal government.[63] In MV the initiative has been implemented in close connection with the Regional Planning Associations (RPAs).[64] The main aim with the Regional Management Initiative, it is argued, is to facilitate economic growth, primarily by mobilizing regional actors and by creating 'concordance' in regional developmental work. Mobilization and unification of regional forces are mentioned as central parts in this work, and the goal, it is said, is to reach consensus on regional developmental issues. The Regional Management Initiatives, so the *Land* government argues, are thus meant to outmanoeuvre 'local egoism' to the benefit of regional partnership.[65]

The aim and the alleged advantages of the Regional Management Initiatives are described in fairly similar terms in all subregions. In Mid-Mecklenburg/Rostock, the Initiative is understood as a response to 'increasing

61 MfABL MV (2001b:24); MfABL MV (2000a).

62 MfABL MV (2001b:2).

63 The initiative is subsided by the fund for 'Verbesserung der regionalen Wirtschaftsstruktur' from within the so-called Common Task *(Gemeinschaftsaufgabe*, GA). At the initial stage, the initiatives were temporary, and subsidies were granted to the end of December 2004. See MfWi MV (2003), *Förderprogramme - Einzigartige Förderkulisse für Investoren in Mecklenburg-Vorpommern*, Schwerin; MfABL MV (2002b), *Raumentwicklung im MV - Initiativen zur Regionalentwicklung*, Schwerin.

64 MfABL MV (2002b); MfABL MV (2004c), 'Regionalmanagement und Regionalmarketing', at *www.am.mv-regierung.de/raumordnung/aktuell_regionalmanagement.html*, 2004-06-25.

65 MfABL MV (2002b).

globalization and Europeanization'.[66] The Regional Manager is assigned to 'bring citizens, the business sector and political forces together, to work in a Private-Public-Partnership-like manner, and to strive for joint efforts within the region'.[67] In West Mecklenburg, it is stressed that the main assignment of the Regional Manager is to 'build up networks and to bring regional actors together'. The leading principle in this work is said to be 'dialogue and cooperation'.[68] The development of 'regional networks' of 'new kinds of regional cooperation' and of 'efficient structures of communication' are understood as central elements in regional developmental work.[69] Also in Western Pomerania, the main task for regional management is understood to be to pick up good ideas, to mobilize regional forces and to work towards a regional consensus.[70]

In Mecklenburg Lake District, the work with regional management could be understood in a somewhat extended context. Just like the other subregions, Mecklenburg Lake District is equipped with its own Regional Manager. The region has, however, also had some success in federal contests such as *Regionen Aktiv – Land gestaltet Zukunft*[71] or *Region der Zukunft*.[72] Owing to this, this subregion has more public-private partnership initiatives than the others. One such example is the so called Regional Development Conferences, described by the RPA as 'a new planning instrument'.[73] The conferences are understood as being bothconferences 'in the usual sense and a continuous, target oriented approach to regional development'. These events aim in brief at 'bringing together communities, local people, businesses, authorities, local politicians and

---

66 RPV Mittleres Mecklenburg/Rostock (2003), 'Regional Management' at *www.rpv-mmr.de/Regionalmanagement.19.0.html#74*, 2005-10-27.

67 RPV Mittleres Mecklenburg/Rostock (2001), 'Regionalmarketing', at *www.rpv-mmr.de/*, 2004-07-01.

68 RPV Westmecklenburg (2004a), 'Grüsswort von Regionalmanager Dr Jörg Köhn', at *www.westmecklenburg-schwerin.de*, 2004-06-25.

69 RPV Westmecklenburg (2004a); RPV Westmecklenburg (2004b), 'Regionalmanagements der Region Westmecklenburg' at *www.westmecklenburg-schwerin.de*, 2004-06-25.

70 RPV Vorpommern (2002), 'Aufgaben des Regionalmanagement – Vorpommern', at *www.rpv-vorpommern.de*, 2004-06-25.

71 This contest was initiated by the Federal Ministry of Consumer Protection, Food and Agriculture with the aim of renewing rural and agricultural districts in Germany. See *www.nova-institut.de/modellregionen/*, 2004-06-25.

72 This contest was initiated in 2001 by the Federal Ministry for Traffic, Building and Housing, and managed by the Federal Office for Building and Regional Planning.

73 RPV Mecklenburgische Seenplatte (2000b), 'Regional Conference', at *www.region-seenplatte.de/html_e/page3.html*.

chambers of commerce in a kind of regional agenda-process'. The working strategy of the conferences is summed up in slogans such as 'process rather than documents' and 'put plans into action!'.[74] According to the Minister for Regional Development, these conferences have been 'exemplary cases' of regional developmental work. In the Minister's mind, such initiatives would lead to 'a unification of regional actors' and to common priorities in developmental work.[75] Moreover, it is stressed that 'private investors, local entrepreneurs and regional and local actors' ought to 'row the same boat' or to 'pull together', and that this particular subregion is a good example in that respect.[76]

Marketing Networks

In all subregions, explicit links are drawn between the Regional Management Initiatives and efforts made for marketing the region. In a majority of the subregions, we encounter the argument that the image of a region and the ability of its actors to cooperate are equally decisive for the region's ability to cope with global competition. In Western Pomerania, the organization *Die Region Vorpommern* is described as a platform for the promotion of the region and as a structure for handling the Regional Management Initiative.[77] In Mid-Mecklenburg/Rostock, the Regional Manager has worked in close cooperation with the regional marketing organization, described as 'an independent non-profit marketing organization ... working on the basis of a Public-Private-Partnership'.[78] In West-Mecklenburg, the regional management function is referred to as the 'interface' between regional planning efforts and efforts made to market the subregion and improve its image.[79] The regional marketing organization is, furthemore, said to work on 'improving the self-confidence and the team spirit within the region'.[80] Efforts at regional planning and marketing initiatives alike are obviously thought to gain from cooperation between public and private partners.

---

74 RPV Mecklenburgische Seenplatte (2000b).

75 Holter, Helmut (2000b), in RPV Mecklenburgische Seenplatte (eds.), *REGIOnal*, No 2, Juni, Neubrandenburg.

76 MfABL MV (2003b), Schwerin; Holter (2000b).

77 MfABL MV (2004c).

78 Region Rostock (2004c) 'Regional Marketing Structure', at *www.regionrostock.de/Marketing_Initiative.267.0.html*, 2004-06-30.

79 RPV Westmecklenburg (2004b).

80 The organization was established in 1999, partly on the initiative of the RPA in this subregion. MfABL MV (2004c).

Cooperation as an Organizational Ideal

The organizational ideal that seems to crystallize in both regions is characterized by partnership, networks, and cooperation among a wide range of regional actors. Thus, a great many initiatives have been taken to increase the level of cooperation within the two regions. The Norrbotten Group, Europaforum, Norrbotten International Council, DW 2020 and the various initiatives for Regional Management could all be understood as institutions embodying a certain organizational ideal. When these initiatives are described in the regional policy discourse, it is stressed that working in a cooperative and partnership-like manner is necessary for promoting the region's best interests.

The accent on cooperation and broad coalitions does not come as a surprise. If the basic sentiment is that the region is in a situation of global competition and that a feeling of regional fellowship and the articulation of a common interest would benefit the region, the bid for cooperation and partnerships is pretty consistent with this. Besides, public-private partnerships, broad coalitions, development alliances and other attempts to unite regional actors around a common development strategy have been regarded as standard elements in today's European Union.[81] The partnership principle is embedded in EU structural funding, in the so-called Lisbon Process and in initiatives such as the Territorial Employment Pacts.[82] Yet initiatives for bringing regional actors together are found not only at the level of the European Union; at the level of the state as well we find the conviction that broad agreements and partnerships will be of use to the region.[83] In a Swedish context, the Regional Growth Agreements (RGA) and later on the Regional Growth Programmes (RGP) prove to be examples of how the central government has launched the idea of partnership. In a German context, federal initiatives such as the aforementioned contests *Regionen der Zukunft* or *Regionen Aktiv* could serve as corresponding

---

81 Deas and Ward (2000:274); Keating (1997:31,36).

82 EC (2000-03-23/24), 'Lisbon Presidency Conclusions', Lisbon, point 37.; Haahr, Jens Henrik (2004), 'Open Co-ordination as Advanced Liberal Government', *Journal of European Public Policy*, 11(2), pp. 209–230; Ahonen, Pertti (2001) *Soft Governance, Agile Union?, An Analysis of the Extensions of Open Coordination in 2000*, Maastricht: European Institute of Public Administration.

83 Finansdepartementet (n.d) *Lissabonstrategin – VM i hållbar tillväxt*, Stockholm; Näringsdepartementet (2002) *Rapport om tillväxtavtalen – andra året*, Ds 2002:34, Stockholm, p.6; Messing, Ulrica (2002a) 'Inledningsanförande' *Tillväxt 2002 – Sverige i framtiden*, Piteå; Messing, Ulrica (2002b), 'Vi vill att hela Sverige ska leva och utvecklas', *NSD, 2002-07-06.*

examples.[84] Partnerships have, as Keating points out, 'come to be almost universally valued'.[85]

Economic and Democratic Benefits?

In the main, public-private partnerships are put forward as a means for creating an innovative environment in the region, favourable to economic growth. Partnerships and broad coalitions are thought to bring trust, social capital and civic spirit to the region. Given that, the mere establishment of a partnership is frequently declared as an improvement in itself. Yet it needs to be said that the hypothesis that regional economic development really benefits from unity and cooperation is very hard to verify. We do indeed find those who claim to demonstrate the huge importance of informal institutions in processes of regional economic development.[86] Still, these claims have been criticized for overemphasizing local factors, neglecting macroeconomic factors and running the risk of reverse causality.[87] It has also been suggested that attempts to explain economic growth in terms of local conditions may result in 'misplaced economics of hope' and in a neglect of other insights and complementary explanations of economic success.[88] It could also be objected that, even if we accept and welcome a scholarly approach to economics in which non-rational or immaterial aspects are recognized, we do not necessarily need to welcome a *policy* which seeks to create or construct these immaterial aspects.[89] Analytical statements simply must not be confused with policy recommendations.

Partnerships are thought to be of advantage not only in improving the region's competitiveness. Working in partnerships and networks is also thought to renew and to revitalize democracy and political life, whereas parliamentary debates and

---

84 Jakubowski, Peter (2001), 'Regionen im Wettbewerb', *Informationen zur Raumordnung*, 8. Bonn: Bundesamt für Bauwesen und Raumordnung.

85 Keating (1997:36).

86 For one of the most influential constributions, see Putnam (1993).

87 Colletis-Wahl, K. and Pecqueur, B. (2001), 'Territories, Development and Specific Resources: What Analytical Framework?', *Regional Studies* 35(5), pp.449-459, p.451; Lovering (1999:384); Das, R. J. (2004), 'Social Capital and Poverty of Wage Labourers: Problems with the Social Capital Theory', *Transactions of the Institute of British Geographers*, 29(1), pp.27–45; Bräuninger, Michael (2002), *Social Capital and Regional Mobility*, Hamburg: Institut für Theoretische Volkwirtshaftslehre, Universität der Bunderswehr, p.3.

88 Amin (2001:1237ff.)

89 Amin (2001).

negotiations are referred to as rigid and old-fashioned in comparison. In a Swedish context, however, regional networks and partnerships have been criticized for being far too inaccessible by the citizens. The actors involved in these partnerships cannot always be held responsible for their decisions in public and general elections, and the prospects of bringing about a public debate about their structure, decisions and priorities have been estimated as limited.[90] On this matter, Amin and Thrift warn against a 'too ready acceptance that the [possible] economic efficiencies generated by networking automatically translate into a gain in equity and democracy'. According to Amin and Thrift, the effect of these networks on democracy is at best unproven. At worst, 'such networks may simply constitute a new weapon in the armouries of the powerful'.[91]

Besides the potential gap between these institutions and the citizens for whom they are thought to work, there is a risk of a power imbalance between and within networks. Amin and Thrift have argued that 'even the "flattest" of contemporary corporate networks are not openly constructed spheres of action, but carefully crafted "heterarchies" of power which distribute tasks, competences and rewards unevenly between members'. It is furthermore argued that these networks are far from being democratic institutions, and that they fail to 'see as their mission to put social goals before profit-based goals'.[92] Keating has described 'development coalitions' as cross-class alliances made up of social and political actors who have all dedicated their activities to economic growth in a specific location. Like Amin and Thrift, Keating points out that these special-purpose agencies have the potential to narrow the development agenda and to neglect social dimensions and tensions within the region.[93]

Cooperation as Community Building

The decision to work through partnerships, networks and development coalitions is not devoid of value judgements, but is linked to a set of ideas about what political work should strive at, and how politics should be executed.[94] Besides,

---

90 Hudson, Christine and Rönnblom, Malin (2003) 'Uteslutande partnerskap?' in Håkan Brynielsson (ed.) *På jakt efter en ny regional samhällsordning!*, Stockholm: Svenska Kommunförbundet; Gidlund (1998:19).

91 Amin, Ash and Thrift, Nigel (1995), 'Institutional Issues for the European Regions', *Economy and Society*, 24(1), pp.41-66, p.42.

92 Amin and Thrift (1995:58).

93 Keating (1998:144-146).

94 Hudson and Rönnblom (2003).

the organizational ideal referred to above forms an important part of both regionalism and of a region-building process. The formation of new partnerships and networks is an example of region-building, not only in the formal, state-building-like sense.[95] As an alternative, it is region-building also in the sense of community building, primarily as these institutions bring the idea of regional solidarity and belonging to the fore.

The formation of identities and feelings of belonging is a long-term process. Therefore, it may be too early to draw conclusions about the effects of all these networks and partnerships on the participants' feelings of regional belonging. It has indeed been suggested that institutional development and change do have implications for territorial identity-formation. Lecours, for instance, has contended that the existence of territorial identities requires an institutionalist or political explanation rather than an explanation that focuses on cultural or structural preconditions.[96] Institutional development and change is said to 'trigger mechanisms of territorial identity formation not only through direct boundary setting, but also by shaping elite relationships, competition and behaviour'. In fact, Lecours even suggests that 'political institutions exercise an independent effect on territorial identities by shaping or even creating political communities'.[97]

The influence of institutions on people's perceptions has been brought out before. Brubaker suggest that the 'institutional arrangements which an individual inhabits' can influence the 'identity, interests and goals' of the individual.[98] Brown has, quite similarly, argued that those who are part of a certain institution are 'under pressure to ... adopt the values, interests and demeanour which are embedded in the institutions which they enter'. This implies, as Brown suggests, that the institutional interests – which in our context is to work jointly for the sake of the region – 'may well come to dominate the interests and values (of those) who occupy the institution'. Individuals who regard themselves as representatives of the working class, of liberal ideas, of women or of some kind

95 For a qualification of the distinction between state-building and nation- or community-building, see Chapter 1.

96 Lecours, André (2001), 'Political Institutions, Elites and Territorial Identity Formation in Belgium', *National Identities*, 3(1), pp.51-68 p.52.

97 Lecours (2001:54).

98 Brubaker (1997:24); see also Brown, David (2000) *Contemporary Nationalism: Civic, Ethno-cultural and Multicultural Politics*, London and New York: Routledge p.21

of minority group when entering an institution, may well change their dispositions and be more focused on joint efforts for the region's best interests.[99]

The partnerships and networks referred to above are all permeated by territoriality. Hence, their single ambition is to promote and support a certain territorial unit. This may imply that the *institutional* socialization provided by these networks can have implications for the *spatial* socialization of their members.[100] That is, the networks may influence not only the members' 'values, interests and demeanour'[101] but also their *socio-spatial consciousness*[102] and the ways in which they peceive a certain territory.

A central point here, then, is that elites in the region may undergo a spatial socialization process that differs from the corresponding process among the public at large. For those who are very much involved in an institution that aims at promoting a certain territorial unit, the territorial unit as such may stand out as more natural, more coherent or more worthy of defence than it does for the public at large. The partnerships and networks could thus provide its members with a socio-spatial consciousness and 'a set of boundary-markers'[103] that are shared by those involved in the institutions in question, but not necessarily by the general public. What we have then is a partial process of region-building that at the present point in time must be understood to influence the policy-making elite to a much greater extent than it affects the grass roots.

There are certainly methodological difficulties in coming to a conclusion about the existence and development of collective identities and feelings of territorial belonging. Because of that, we cannot yet take these networks' and partnerships' real effects on people's identities for granted.[104] Previous studies of the effects of regional political activity and of the establishment of regional political institutions on regional conciousness have in fact failed to establish any causal relation between these factors.[105] Moreover, few people are socialized into membership of just one spatial entity; and the fact of *regional* institutions insisting on *regional* solidarity does not exclude the existence of other institutions, insisting on other kinds of solidarity with greater success. Attempts

---

99 Brown (2000:36).

100 Paasi describes *spatial socialization* as 'the process through which actors and collectives are socialized as members of certain territorially bounded spatial entities'. Paasi (1996a:8).

101 Brown (2000:36).

102 Paasi (1996a:65).

103 Brubaker (1997:24).

104 Breuilly (1994:277).

105 Keating and Loughlin (2002:5).

to strengthen a particular territorial frame of reference can furthermore be counter-productive. People may, as Breuilly suggests, get fed up with what they perceive as exaggerated propaganda.[106] Finally, it could also be objected that feelings of regional belonging can flourish, not only because they are institutionalized by an elite but also because of their suppression.[107]

The Regional Elite as a Thought Collective

Despite the methodological problems in coming to conclusions about the existence of, and reasons for, people's territorial identities, I maintain that the organizational ideals that emerge here are important parts of both regionalism and of region-building. Politicians from all parties prove to be very keen to stress the importance of uniting around a common regional interest. To work for the region's best interest through participating in partnerships and broad alliances is an organizational ideal endorsed by many. The idea of 'the region' and the feasibility of uniting around a common regional interest appear, therefore, to be very consistent for a great share of the regional policy-makers.

Even if 'territorial frames of reference'[108], identities or feelings of belonging rarely change swiftly, some actors evidently host an *outspoken ambition* to change the geo-political frames of reference. By taking part in the regional institutions mentioned, the members seem to 'learn' that the region *should* be the proper geographical level of solidarity. In this matter, Fleck's idea about *thought collectives* actually proves to be very useful.[109] A thought collective is, in Fleck's sense, a group of actors involved in mutual thought-related interaction.[110] By means of an intense exchange of ideas and thoughts, a so-called *thought-style* is developed, which makes out of individual thoughts a common product. Here, the idea that the region (as a whole) would gain from extensive networking could be mentioned as one example of a thought-style.[111] This thought-style can, in turn, be understood as a result of a particular institutional framework or thought collective where several regional actors take part and where the state, the European Union and – seemingly – a number of academics are influential

106 Breuilly (1994:277); Keating and Loughlin (2002:6).

107 Brown (2000:21,32-33).

108 Keating and Loughlin (2002:6).

109 See Fleck, Ludwik (1997), *Uppkomsten och utvecklingen av ett vetenskapligt faktum*. Stockholm/Stehag: Symposion.

110 Fleck (1997:110).

111 Fleck (1997:122).

discussants. The problematic aspect of this development is that regional representatives, by reproducing the idea of the region as a coherent space in terms of interests and ideologies, run the risk of neglecting the presence of diverging interests within the region. Political affiliations, ethnicity, class, gender or other bases for solidarity might be set aside for the sake of the region.

Rationalized Myths and Symbolic Power

Working in partnerships and in broad alliances is often referred to as the best and the most rational and appropriate ways to organize political work at the regional level. Whether this assessment is right or wrong is, however, hard to judge. Rationality, reason and suitability are far from being 'absolute' or free from subjectivity and ideological conviction. Accordingly, the aforementioned organizational ideal could be understood as a set of *norms* of rationality or as rationalized myths rather than as an instance of 'pure' rationality.[112] The myths about the excellence of partnerships in regional policy-making are, however, not produced by these regions in isolation. Instead, they are produced and reproduced in interaction with academics, with the state and with the European Union. The European Union includes partnerships among its conditions for the receipt of structural funding, and the Swedish and German states respectively do encourage partnership-like working methods. Still, the regional level is not only adjusting to these demands, but is contributing just as much to the reproduction of them. When regional representatives form partnerships and networks, they not only accept the rules of the game. They also *confirm* the suggested rationality of this way of organizing political work. This is particularly the case not only when regional representatives have to form partnerships, but also when they are free to choose whatever organizational form they like.

Many of the networks and partnerships formed at the regional level are based on voluntarism. Most of them do not change the formal tasks of the participating parties and, in many cases, they possess a symbolic rather than an executive power.[113] This is, however, not to say that these networks, development coalitions or partnerships are insignificant. To the contrary, they prove to be important institutions in the regional organizational landscape.[114] These

112 Johansson, Roine (2002), *Nyinstitutionalismen inom organisationsanalysen*, Lund: Studentlitteratur, p.30.

113 Bourdieu (1991).

114 Süssner, Josefina (2002), 'Nyregionalismens organisatoriska landskap' *Nordisk Samhällsgeografisk Tidskrift*, 35, pp. 5-23.

partnerships constitute, particularly in a Swedish context, an important part of the base on which the political debate about regional development takes place. These institutions rest, furthermore, upon a normative basis, where a certain enthusiasm towards intra-regional cooperation, common regional stances and a high level of interaction in the European policy process are common sense. Accordingly, alternative agendas and visions will have severe problems with having any impact.

The suggestion is that networks, groups and partnerships of the kind referred to here could be understood as rites and practices through which a certain ideological conviction is manifested. The importance of rites and practices in ideology has been elaborated in Althusser's works on ideological state apparatuses.[115] Althusser's understanding of ideology differs indeed from the way in which ideology is understood here. Yet Althusser's notion of how institutions affect political beliefs and convictions can help us increase our understanding of the role of partnerships and networks in regionalism and region-building. From Althusser's perspective, political activities function as rites through which beliefs and ideological convictions are manifested. It is, according to Althusser, not necessarily the ideology that determines the ceremony or the rite.[116] The ceremony and the rite may, as an alternative, combine so as to strengthen the ideology and its normative foundations. Here, I would suggest that participation in various kinds of 'development coalitions'[117] can be understood as ceremonies that strengthen the conviction that the region, unity and economic growth are fundamental values that need to be defended. Even if these institutions have a limited formal power, they embody by their mere existence the idea of the region as a coherent sphere of interests.[118] Accordingly, these institutions can be understood to contribute to a process of regional community-building.

115 Althusser, Louis (1984), *Essays on Ideology*. London: Verso.
116 Althusser (1984).
117 Keating (1998:144).
118 Süssner (2002).

7. Means of Regional Mobilization

In nationalist and regionalist ideologies alike, notions about culture, history and 'belonging' are among the features used for mobilizing ends. The role of culture and identity in political life in Norrbotten and Mecklenburg–Western Pomerania is still a very open question. In both regions, the world-view communicated and the core values and organizational principles adopted are closely connected to new regionalist ideals. Yet the new regionalist' position on matters like culture and identity is not completely settled. Previous research has suggested that the new European regionalism is modernizing, cosmopolitan and forward-looking.[1] In that sense, it has been contrasted to the regionalism of the 1960s and 1970s, which has occasionally been characterized in terms of 'ethno-nationalistic movements' with 'the right to roots' an important part of their agenda.[2]

Still, there are many indications that culture and identity do play a crucial role also in recent versions of European regionalism. In research as well as in politics, it is claimed that the need for regional identities is increasing due to globalization and European integration.[3] Regional identity and culture are furthermore thought to facilitate regional economic development in a variety of ways.[4] Below, I will inquire into how culture and identity have been represented within the two regions.[5] Are these features primarily referred to as 'something that is continually being created and recreated' or are they treated as something fixed, inherited and deeply rooted in the history of the region?[6]

---

1 Keating (1998:73); Harvie (1994:3).

2 Schmitt-Egner (2002:188); Gren (1999:31).

3 Castensson (1994:51); Petersson and Clark (2003).

4 Kockel, Ulrich (2002), *Culture and Economy: Contemporary Perspectives*. Aldershot: Ashgate; Power, Dominique (2002), '"Cultural Industries" in Sweden: An Assessment of their Place in the Swedish Economy', *Economic Geography*, 78(2), pp.103-127, p.105; Gregson, Nicky, Simonsen, Kirsten and Vaiou, Dina (2001), 'Whose Economy for Whose Culture?', *Antipode*, 33(4), pp. 616-646.

5 Hylland-Eriksen, Thomas (1999), *Kulturterrorismen*, Nora: Nya Doxa, p.23; Keating *et al.* (2003:37); Gellner (1997:19).

6 Keating (1998:155).

Norrbottnic culture is recurrently described as the result of long-term historical processes. In the Cultural Policy Programme (CPP) from 1994, it is said that 'our origin is rooted in nature, climate, working life and in the conditions under which we live'.[7] In other contexts, too, it is emphasized that 'nature, climate ... and our geographical position' have moulded Norrbottnic culture in many ways. It is also maintained that these features continue to influence social and cultural life in Norrbotten as strongly today as in the past. [8]

As well as nature and climate, the economic history of the region is said to have moulded Norrbottnic cultural life. Societal development in the entire Barents region, it is argued, has been characterized by 'a colonial perspective and by having exploitation of raw materials as the main driving force'.[9] According to the County Council, Norrbotten has always been 'a pioneers' area', and a place to which 'people came from the whole country with different cultures'. Accounts of these groups of immigrants include references to 'huntsmen, settlers ... navvies, woodmen, and ironworkers'. The influences from these groups, it is argued, make 'identity- and border crossing to a part of our cultural pattern'.[10]

It is often maintained that there is a certain *mentality* in Norrbotten. In some contexts, references are even made to 'the Arctic Man' and to the climatic and cultural conditions that formed him.[11] We learn that the 'Arctic Man' has a particular 'sense of beauty' and a particular 'attitude towards life and to the world around'.[12] In other contexts, the 'Norrbothnian' is said to 'love . . . nature and outdoor life of all kinds'[13], and to value freedom, environment and 'the fluctuations of the seasons' particularly highly.[14] In *The Guide*, already much referred to in previous chapters, we learn that 'social intercourse amongst us is easy and free from idle words; we are straightforward and honest, explicit and easy to get in touch with; we are loyal and willing to work – all of which are characteristics that we certainly do not need to be ashamed of'.[15] One County

7 NLL (1994a:15).
8 NLL (1994a:15, 22); NLL (1996b:5).
9 NLL (1996b:4).
10 NLL (2000a:2; NLL (1994a:3,16).).
11 NLL (1996b:5).
12 NLL (1994a:15).
13 Lst Norrbotten (1999a), *Norrbotten – Mångfald och Kontraster*, Luleå.
14 Lst Norrbotten (1999a); NLL (1994a:15).
15 Partnerskap för Norrbotten (1995:12).

Governor holds, in much the same manner, that 'Norrbothnians are known for being straightforward' and that 'no one needs to doubt the Norrbothnians' generosity, stability, and reliability'.[16] A later County Governor maintains that 'we are open, generous and very straightforward'. The Norrbothnians are easy to communicate with, it is held, and this 'makes it easy to live in the County of Norrbotten'.[17] In yet other contexts, it is stressed that the Norrbottnic workforce is both motivated and loyal, and that 'we can be proud and happy with being Norrbothnians'.[18] A representative for the Liberal Party maintains in this matter that 'Norrbothnians are a people of a certain kind'.[19]

The indigenous minority populations in the region are said to have had a considerable impact on Norrbottnic culture and identity. Norrbottnic cultural life is repeatedly referred to as 'multicultural' and as marked by an 'interplay' between Swedish, Saami and Finnish influences.[20] In a variety of publications from the County Council, it is stressed that 'we have a multicultural identity' and that social progress in Norrbotten historically has been 'clearly influenced by the culture of the indigenous groups'.[21] In the CPP, we learn that 'our Norrbottnic identity is closely connected to the identity that the indigenous people ... created through their interplay with nature and climate and through their work and general living conditions'.[22] In a later publication from the County Council, it is held that Norrbottnic identity is 'strongly affected by the languages used here: Swedish, the river valley dialects, *Meänkieli*, Finnish and Saami'.[23] The then County Governor, however, refers to the Saami, the Swedish and the Finnish as 'the three cultures of the region'.[24] In the CPP, too, these three cultures are said to constitute 'the essence of our cultural heritage'.[25] On the whole, Norrbottnic culture is described as the unified result of a rather uncomplicated interplay

---

16 Rosengren, Björn (2000), 'Framåt för Norrbotten' at *www.norrbotten.se/arkiv/artiklar/framot.htm.*

17 Eriksson, P.O (2003a), 'Företagandets, samarbetets och möjligheternas län', in Lst Norrbotten (ed.), *Bilder av Norrbotten*, Luleå, p.4.

18 Norrbottengruppen (1999:24); Lst Norrbotten (2004b), *Kompetensfrågor i samverkan – ett magasin om kompetensförsörjning*, Luleå, p.3.

19 Grönlund, Anna (2002), 'Jag vill måla om bilden av Norrbotten', *NSD*, 2002-10-10.

20 Lst Norrbotten and Norrbottens Museum (1998), *Kulturmiljövårdens idé*, Luleå, p.1.

21 NLL (1994a:3) NLL (1996b:4).

22 NLL (1994a:15).

23 NLL (2000a:2).

24 Marklund, K. (2001a), 'Landshövdingen hälsar välkommen', at *www.bd.lst.se*, 2001-11-23.

25 NLL (1994a:21).

between three distinct cultural entities. This interplay is seen as valuable and as worthy of protection. In *The Guide* it is suggested that the establishment of a Saami cultural centre would 'stimulate culture based on our heritage'.[26] Likewise, in a later publication, the work on preserving Saami and Finnish language and culture is represented, not only as a matter for the indigenous groups themselves, but as 'a concern for the entire region'.[27] In the CPP, it is maintained that 'the preservation of the Finnish and Saami languages is a fundamental condition for our multicultural heritage'.[28] In the 2004–2007 Regional Development Program, it is once again stressed that the specific multicultural character of the region must be preserved.[29]

Why is Norrbottnic Culture supported?

Even if we are told that there *is* a common culture and identity within the region, it is also suggested that the region and its inhabitants would gain a lot from a strengthening of this very culture and identity. *The Guide* proposes that knowledge about 'the attitudes and the mental structure of the Norrbothnians' needs to be deepened. It also makes out a need for 'textbooks and teaching aids about our history'.[30] In several publications, the County Council asserts that one of its ambitions in the field of cultural and media policy is to work for a reinforced regional identity.[31] The County Administrative Board holds that the cultural heritage and knowledge about it is 'the basis of our identity'. In consequence, it is argued, the cultural heritage should be preserved and used for increasing knowledge about 'us and our identity as Norrbothnians'.[32] In 1999, the County Administrative Board financed the publication of a junior-level school textbook about Norrbotten. The book was referred to as 'a long-term investment', aiming at 'increasing childrens' awareness and knowledge about their own county'.[33] In other contexts, local politicians come up with similar

26 Partnerskap för Norrbotten (1995:14).
27 NLL (1996b:5).
28 NLL (1994a:21).
29 NLL (2004b) 'Om Norrbottens Regionala Tillväxtavtal', at *www.nll.se/hg4.aspx?id=13985*, 2004-05-19.
30 Partnerskap för Norrbotten (1995:14).
31 NLL (1994a:6,11); NLL (1996a:19); NLL (1998c:32); NLL (2000b:35).
32 Lst Norrbotten and Norrbottens Museum (1998:1).
33 Lst Norrbotten (1999b), 'Min lilla bok om Norrbotten', *Pressmeddelande*, Luleå.

notions by arguing that there is a need for efforts to preserve the 'rich but neglected history' of the region.[34]

A strengthening of regional culture and identity is presumed to serve several interrelated but yet economic ends.[35] *First*, a lively cultural life and a distinct profile are said to be of advantage in drives to attract tourists, investors and other outsiders to the region. *Second*, a firm attachment to the region and a strong regional identity are presumed to counteract negative migration flows and to benefit intra-regional mobilization. Culture and identity are also thought to give strength and stability to the region and its inhabitants in the age of globalization. *Third*, culture and cultural heritage are, to an increasing extent, understood as a line of business in their own right. Below, we will see how these three interpretations are manifested in the regional policy discourse.

Culture and Image as a Way to Attract Interest

Regions are – as emphasized in Chapter 4 – mainly understood as competitors in a global race for economic growth. In this context, a rich and lively cultural life is perceived as a comparative advantage. A distinct regional culture and an 'atmosphere' of a certain kind are seen as beneficial in campaigns to attract investors, companies and workers to the region. These notions are found in several political camps. The Social Democratic Party in the region stresses that culture not only is 'an important part of the heart and soul of a region', but also an 'important factor when it comes to the establishment of companies and economic growth in the county'.[36] A representative for the Left Party holds that culture is a key element if a region is to be attractive or interesting for investors. 'If we turn our region to a nicer, more exciting ... and more attractive place', it is maintained, 'external actors, investors, companies and people would be more willing to settle here'.[37] The County Council maintains in its CPP that 'recent research has revealed the great importance cultural life has for the establishment

---

34 Eriksson, P. *et al.* (2004), 'Vi är stolta over Fattenborg', *NSD*, 2004-04-21.

35 NLL (1994a:5-6); NLL (1994b:22); NLL (1995:12); NLL (1996a:17); NLL (1997c:25); NLL (2004a), 'Statliga pengar till kultur och infrastruktur', *Pressmeddelande 2004-03-29*, Luleå; Petersen, Karl (2001a), 'Kämpa för Norrbotten', *NSD*, 2001-11-06.

36 SAP Norrbotten (2004b), 'Tillväxt och växtkraft i hela Norrbotten', *Uttalande från DS -kongressen 2004-07-27/28*, Norrbotten; Thörnlund, Lennart and Henriksson, Gerda (2002), 'Kultur är utvecklingskraft', *NSD*, 2002-03-05.

37 Lööw, B. (2004), 'Så lockar vi investerare till Luleå', *NSD*, 2004-09-06.

of companies and for the recruitment of well-educated workforce'.[38] In the 2004-2007 Regional Growth Programme for Norrbotten, culture is mentioned as an important element in attempts to make the region more attractive.[39]

The idea that a rich cultural life would increase external interest in the region is closely connected to a discussion about *images*. The image of a region is thought to be of crucial importance for its economic performance. To judge from the policy discourse, an image can be either true or false – either positive or negative. In *The Guide*, we learn that there are 'a great many negative myths about Norrbotten and the Norrbothnians'.[40] These myths or fallacies are seen as a serious problem; and in order to overcome them it is suggested, in both *The Guide* and in a number of other contexts, 'we need to get better at spreading a true picture of Norrbotten and ourselves'.[41] This allegedly true or fair picture must, according to *The Guide*, 'start out from the present time and not be burdened by clichés from times long past'.[42] In several documents from the late 1990s, we find calls for true and updated images repeated almost down to the last letter.[43] In a number of other settings, it is stressed that Norrbotten needs to 'improve its trademark' and that the cultural institutions in the region should 'try to obtain a distinctive regional profile'.[44]

The conviction that a 'true' and positive image is of crucial importance for the regional economic development led to the decision to initiate a number of image campaigns during the late 1990s. *The Guide* was, as pointed out earlier, one of the first attempts to bring a large number of regional actors together around a common strategy for regional development. The work preparing and following up *The Guide* was succeeded by the 'Information Project'[45], initiated in

---

38 NLL (1994a:29).

39 NLL (2004b), 2004-05-19; See also Gröning-Degerlund, Lotta (2002), 'Det behövs mer än ett konserthus', *NSD*, 2002-08-17.

40 Partnerskap för Norrbotten (1995:12).

41 Partnerskap för Norrbotten (1995:12); SAP Norrbotten (1997:3); Holmqvist, (1998a:3); Grönlund (2002); Gröning-Degerlund (2003a), 'Ta befälet, Eriksson', *NSD*, 2003-03-14; Lindström, Olle (2001), 'Ett nytt vägval krävs för att göra länet attraktivare', *NSD*, 2001-11-27; Eriksson, P.O (2005), *Sprid inte Myter!*, Lst Norrbotten: Luleå, p.1.

42 Partnerskap för Norrbotten (1995:12).

43 NLL (1996c:5); NLL (1994b:22); NLL (1995:12); NLL (1996a:13).

44 Partnerskap för Norrbotten (1995:12); NLL (1999:31); NLL (1994a:6,9); NLL (1994b:23); NLL (1995:14); NLL (1996a:18); NLL (1997c:25); NLL (1998c:33); NLL (2000a:4); Framtidslandet (2004), 'Intervju med Kari Marklund', at *www.framtidslandet.nu/index.asp*, 2004-05-19.

45 *Informationsprojektet* in the original.

November 1997 and completed in October 2000. This project was supported by representatives from regional trade and industry, by the municipalities, and by the so-called Norrbotten Group (see Chapter 6). The project aimed at 'making the Norrbothnians even more proud of their county', to spread 'a true image of Norrbotten', to 'strengthen the positive image of Norrbotten', to 'strengthen regional identity' and to 'project Norrbotten as a county of contrasts, multiculturalism and excitement'.[46]

Having completed the Information Project, the same group of actors united around the project 'Norrbotten Presents'.[47] One part of that particular project was the so-called Norrbotten Days[48] – an event marketing initiative promoting the county in Stockholm, Malmö and Gothenburg. Norrbotten Days was supposed to increase interest in the region in general terms, to 'spread knowledge about Norrbotten as a progressive and positive region'[49], and to spread general information about tourist attractions, housing, educational programmes, working life and business opportunities in the region.[50] It was also maintained that the event made it possible for a variety of representatives of the region to stand 'side by side and demonstrate a unity and a willingness to cooperate'.[51]

Internal Mobilization and Inner Orientation

A proper and positive image is, however, regarded as being important not only for efforts to attract outside actors to the region. It is also said to be crucial for the region's inhabitants. In order to be attractive to others, it is argued in *The Guide*, the region and its inhabitants need to be more proud of their own culture and identity. The target group in promotions of the region is therefore not only outside 'individuals, firms and investors'. On the contrary, it is maintained, 'the first target group that needs to be canvassed is we'. For these reasons, it is

---

46 Vikström, Inger (2000), *Slutrapport, Informationsprojektet*, Luleå: Lst Norrbotten; Lst Norrbotten (2000), 'Välkommen till Länsstyrelsens informationsprojekt om Norrbotten', at *www.norrbotten.se/infoproj/intro.htm.*

47 *Norrbotten Presenterar* in the original.

48 *Norrbottendagarna* in the original.

49 NSD (2003b) 'Landshövdingen vill ge en positiv bild av länet', 2003-06-14.

50 Övertorneå Kommun (2002), 'Övertorneå i Göteborg' at *www.overtornea.se/nyheter/goteborg.shtml, 2004-05-19*; Haparanda Kommun (2002), 'Många Tornedalingar på Gärdet', at *www.haparanda.se/haparandaklubben/gardet2003.htm*, 2004-05-19.

51 Lindgren, Ulf and Riklund, Tomas (2004), *Norrbotten presenterar, slutrapport*, Luleå.

argued, 'we Norrbothnians need to learn more about our own county ... to get better at identifying our advantages ... and to decide what image we want to pass on'.[52]

The County Council's Media Policy Programme states that media policy in the region should strive to 'make us, other individuals, companies and investors aware that Norrbotten is a region with qualities, a region 'worthy to be reckoned with'.[53] The Council also maintains that 'Norrbothnians, cultural minorities ... and local initiatives' should 'get the opportunity to make their voices heard' in the media.[54] Therefore, it is argued in a number of documents, regional culture and media policy should strengthen regional identity and 'reflect the diversity and distinctive features of the region'.[55]

One recurring argument in this respect is that cultural affiliation and feelings of regional belonging would make people take more responsibility for the development of their region.[56] Regional culture and identity 'based on what's uniting us and our distinctive features' are said to weld people together, to contribute to building pride and self-esteem in the region, and to act as a 'uniting bond between people'.[57] A Norrbottnic County Governor maintains that northern Italy and other 'sucessful' regions are distinguished by 'strong social cohesion, a strong regional culture, a strong local and regional identity'. This, it is argued, amounts to 'strong social capital' and is important if the region is to develop.[58] According to the County Council, culture has the capacity to stimulate both self-confidence and identity, and to have a positive impact on regional economic development.[59] Arts and culture, it is argued, are therefore of strategic importance in the struggle against depopulation and economic stagnation.[60]

---

52 Partnerskap för Norrbotten (1995:10-14).

53 NLL (1996c:4-5).

54 NLL (1996c:3,8); NLL (1996b:10); NLL (1998b) *Kultur ger livskraft och ett starkare Norrbotten*, Luleå, p. 4

55 Henriksson (1998:3); NLL (1996b:10f); NLL (1996c:3,6-7); NLL (2000a:5); NLL (1996a:19); NLL (1998c:32); NLL (1999:31).

56 Hedström, J.O and Westerberg, J.O (1992), 'Förord', Lundin, Kerstin (ed.), *Norrbottens synliga historia: Norrbottens kulturmiljöprogram*, Luleå: Lst Norrbotten, p.5; Framtidslandet (2004); SAP Norrbotten (2002:3).

57 Hedström and Westerberg (1992:5); NLL (2000a:2); Petersen, Karl (2001a); NLL (1996c:7) SAP Norrbotten (2001:13).

58 Marklund (2002).

59 NLL (1996b:7); Henriksson (1998:2); NLL (1994a:3); See also Vänsterpartiet Norrbotten (2001), *Motion om översyn av den samlade kulturbudgeten*, Dnr 774-00, Luleå.

60 NLL (1994a:17,29).

Culture is thought to influence the 'mental climate' of the region, to contribute to 'vigorous trade and industry' and to be 'permanently positioned as one of the driving forces behind economic growth'.[61] With that, investments in culture are seen as a 'necessary prerequisite for a growth-friendly climate in the county'.[62] Another recurring argument is that internationalization implies an increasing need for firm regional identities. In *The Guide*, we learn that 'it is important to belong to something that one is proud of, particularly when national borders are loosened up by internationalization'.[63] People from all parts of the region are called upon to 'take pride in the fact that they are both important and part of a greater community'. This community, it is contended, 'may not stick together through thick and thin'[64], but must stand united 'when a storm is brewing'[65].[66]

Several voices contend, as was argued in Chapter 4, that 'structural transformations' and 'extensive migration' have resulted in a 'new', 'borderless' and 'competitive' Europe.[67] These changes have, according to the County Council, weakened the nation state and made a 'strong cultural identity' into 'a key issue for modern Norrbothnians' soul and self-esteem'.[68] The 'need for a regional identity' is said to be increasing 'among both immigrants and residents', and culture is said to 'build up our self-esteem' and to provide Norrbotten with 'a more distinct profile, internally as well as externally'.[69] This self-esteem and cultural identity, it is held, will be 'needed in the contacts with others outside our county'.[70]

Culture as a Line of Business

Apart from being interpreted as an advantage for social life, for creativity and for the image of the region, culture is also understood as being a line of business in its own right.[71] Cultural institutions subsidized by the public sector should

---

61 Norrbottengruppen (1999); NLL (1996b:8).
62 NLL (1994a:29); Thörnlund and Henriksson (2002).
63 Partnerskap för Norrbotten (1995:12).
64 *I vått och torrt* in the original.
65 *När det börjar blåsa* in the original.
66 Partnerskap för Norrbotten (1995:14).
67 NLL (1994a:3,9,29); NLL (1996b:7); NLL (1998b:2); Thörnlund and Henriksson (2002).
68 NLL (1994a:3,9); NLL (1996b:7).
69 NLL (1994a:9,29); NLL (1996b:7); NLL (1998b:2).
70 NLL (1994a:29); NLL (1996c:4).
71 SAP Norrbotten (2002:3).

indeed be 'of regional character', and yet support for cultural initiatives from the County Council should 'strengthen the Norrbothnians' regional identity', 'create diversity, and widen perspectives'. According to the County Council, however, regional cultural policy should also 'contribute to regional development', 'promote cultural industry', 'contribute to the regional economy' and 'increase employment'.[72]

The decline of the industrial sector is in many cases thought to give the tourism sector in general, and cultural tourism in particular, a more prominent role in the Norrbottnic economy.[73] The Centre Party emphasizes that 'the exposure of our distinctive character ... is an important factor for the tourist trade'.[74] The Liberal Party stresses in much the same manner that the 'distinctiveness of Norrbottnic culture needs to be developed and made more accessible for tourists'.[75] In the Regional Growth Agreement, culture and the media are conceived as 'ever stronger driving forces for employment and growth'. Increased cooperation between 'the cultural sphere, the private sector and the tourist trade' is, furthermore, thought to bring economic growth and development to the region.[76]

New Cultures, New Identities?

The ambition to give the region a distinctive profile implies the accentuation of some characteristics and the omission of some others. In *The Guide*, 'exotic strains' are explicitly described as worthy of value, both for the regional inhabitants themselves and in interaction with 'the outside world'. Norrbotten, it is suggested, 'has a great and largely unexploited asset in its distinctive character'. There is, according to *The Guide*, an 'obvious enticement in the exotic' that the region can make even more of. It is also contended that 'we' must get better at 'taking care of our exceptionality, of those things that are unique for us, of those things that cannot be copied'. In order to cope with that, it

---

72 NLL (1994a:6, 11); NLL (2000a:4). See also Eriksson, S. *et al.* (2004).

73 Eriksson, S., Olsson, L., Lennartson, L., Agnemar, A. and Svensson, R. (2004), 'Turismen blir allt viktigare för de små kommunerna', *NSD* 2004-09-06; Eriksson, P. *et al.* (2004); Partnerskap för Norrbotten (1995:28,34); NLL (2004a); Lst Norrbotten and Norrbottens Museum (1998:1).

74 Centerpartiet Norrbotten (1998:16).

75 Folkpartiet Liberalerna Norrbotten (1998:4).

76 Norrbottengruppen (1999).

is said, 'we must learn to see the value in these things ourselves, and to be proud of what distinguishes us from others'.[77]

The importance of spreading knowledge about the so-called 'specificities' and unique features of the region are repeated elsewhere.[78] The Centre Party in the region stresses that it attached 'great value to the preservation of our cultural distinctiveness'.[79] In the Cultural Policy Programme, it is maintained that Norrbotten is seen by others as special, and that this is something that the region needs to make the most of. According to the programme, 'the stock of the authentic is rising'.[80] Therefore, it is argued, 'it might not be that bad to be special?'.[81] In the Media Policy Programme, it is stressed that 'we need to get better at emphasizing our regional identity, our way of living, reasoning, thinking and acting'. Moreover, it is declared, 'we need to stress the geographical, cultural, climatic and other peculiarities that distinguish Norrbotten from other regions'.[82] In a publication from the County Administrative Board and the Norrbotten Museum, it is stressed that 'tourists want to experience, see and face the authentic'.[83]

As the distinctive, the authentic and the traditional are thought to be of particular advantage for regional economic development, we may ask how 'new', immigrant or alternative cultures are represented in the regional policy discourse. Basically, people from Norrbotten are thought to have fewer problems than the average in meeting 'new cultures'. In the CPP it is maintained that the 'historical heritage of interplay between three cultures – the Saami, the Finnish and the Swedish' gives the region 'an advantage when meeting new cultures and when developing cultural diversity'. The multicultural character of the region is held to be an advantage as Norrbotten has become 'a county of Latin Americans, Iranians, ex-Yugoslavs, and people from many parts of the world'.[84] A later publication from the County Council holds that culture is 'the glue that keeps society together', but that the multicultural past of the region also gives Norrbotten a comparative advantage over other regions when it comes to 'meeting people from other cultures'.[85]

---

77 Partnerskap för Norrbotten (1995:10).
78 Rusiniemi (2002:30); Marklund (2003).
79 Centerpartiet Norrbotten (1998:4).
80 *'ökat i kurs'* in the original.
81 NLL (1994a:15).
82 NLL (1996c:5).
83 Lst Norrbotten and Norrbottens Museum (1998:1).
84 NLL (1994a:3) see also NLL (1998b:7).
85 NLL (1998b:7).

Parts of the discussion on identity and culture in MV revolves around the fact that the *Land* is a new political construct made up of two units – Mecklenburg and Western Pomerania.[86] In the MV constitution, it is laid down that cultural policy must care for the particular interests of both parts of the *Land*.[87] In the parliamentary address from 1998, the SPD President states that 'our *Land* consists of two parts – Mecklenburg and Western Pomerania – brought together as a *Bundesland* only after *die Wende*'. The German word *Bindestrichland*[88] means a *Land* that is constituted by two geographical units, coupled together through a change of political borders.[89] While referring to *Bindestrichländer* in former West Germany, the President argues that the integration process and 'the growing together' has functioned at its best when both parts of the *Land* have been 'aware of their own history and their own identity'.[90] In a later publication, however, the government maintains that many *Bindestrichländer* have had 'problems with finding and developing a common civic identity and tradition'.[91] In yet other contexts, it has been stressed that culture and cultural heritage have been 'important for the growing together of the two parts of the *Land*'.[92] Besides, outside political commentators have argued that the citizens of MV, 'like in every other *Bindestrichland*', are having problems creating an identity, and that it will take time before a *Landesidentität* settles here.[93]

A Common Regional Culture?

Given that Mecklenburg–Western Pomerania is referred to as a *Bindestrichland* of recent formation, we may expect to find few claims that there exists a deep, common and historically derived identity within the borders of the *Land*. Yet we do find voices maintaining the existence of a common regional identity in

86 See Chapter 3.

87 StK MV (2000d), *Verfassung des Landes Mecklenburg-Vorpommern*, Schwerin, art.16.

88 Literally, 'hyphenated *Land*'.

89 Bahra (1997:11); Kuhn (2000:152); Werz and Hennecke (2000:6).

90 StK MV (1998:13f.).

91 StK MV (2000c).

92 MfBWK MV (2002).

93 Lapp, Peter Joachim (2000), 'Ein Land auf der Suche nach seiner Identität', *Aus Politik und Zeitgeschichte*, Nr. 37-38, pp.15-22.

Mecklenburg–Western Pomerania, rooted in history and significant for its inhabitants.

The Christian Democratic Union (CDU) is among those who are most inclined to assert the existence of a common regional identity. According to the CDU group in the *Landtag*, the 'regional identity and the attachment to the *Land*' are now 'clearly pronounced among the inhabitants'.[94] This is promoted as a great advantage. Thus, it is contended, the GDR past implied that people were 'systematically weaned from Christian values, and alienated from their traditional, national and regional roots'.[95] In another CDU publication, people in MV are said to 'connect themselves more and more to their traditional roots' and that this is 'the way in which a strong regional identity emerges'.[96] In the CDU election programme from 2002, it is declared that the 'Mecklenburgers and Western Pomeranians love their *Land*', and that they are 'proud of what they have achieved'.[97] A former CDU President of the *Land* claims, furthermore, that immediately after *Die Wende* 'people sought out their roots and their identity in great euphoria'. At this point in time, however, MV is said to be 'firmly established in its identity and its roots'. This, it is maintained, is a good basis for further political development in the region. Thus, it is argued, '*Heimat* and identity' should not be understood as '*Deutschtümelei*',[98] but as 'the basis from which people start their lives and to which they return in times of distress'.[99]

The CDU is not, however, the only party to bring ideas about cultural roots and identity to the fore. In publications from the SPD/PDS government and from the subregions, it is maintained that MV is 'an old cultural region', among 'the oldest cultural landscapes in Germany', and a *Land* with 'old cultural traditions'.[100] The Hanseatic past and closeness to the Baltic Sea and the Scandinavian countries are said to have been of considerable importance for the cultural development of the *Land*.[101] The cooperation between MV and southern Sweden is said to be 'based upon a piece of common history and culture', and the countries around the Baltic Sea, it is argued, 'behold a common cultural past'

94 CDU Landtagsfraktion MV (2002:37).

95 CDU Landtagsfraktion MV (1996:22).

96 CDU MV (2001).

97 CDU MV (2002:2).

98 'jingoism' or 'a chauvinist and reactionary affinity to Germany'.

99 Seite, Berndt (2002), 'Wohin, Mecklenburg-Vorpommern?', *Die Welt*, 2002-10-02.

100 StK MV (1998:26); StK MV (2002c:30); Region Rostock (2004b), *Kulturtouren*, Rostock.; Kulturforum Mecklenburg-Vorpommern (2002-02-26), *Ohne Titel*, Schwerin, p.1.

101 MfBWK MV (2003); MfBWK MV (2001a), 'Der Kulturaustausch der Völker im Ostseeraum', *Rede des Ministers Kauffold auf Ostseeforum*, Warnemünde, p.2, 7.

that is of importance for contemporary political cooperation.[102] According to the Minister for Cultural Affairs, further contacts and regular meetings between representatives from all the Baltic Sea states will 'promote the cultural identity of this historical region'.[103]

A feature emphasized in many discussions on regional identity and culture is *Plattdeutsch*, *Niederdeutsch* or Low German.[104] Low German has since 1999 had an official status as a regional language, and in both tourist guides and in publications from the government the language is said to be an important regional trait.[105] In a publication from the State Chancellery, we learn that those who want to 'get to know the people of Mecklenburg and Western Pomerania should listen to *Plattdeutsch*'. The regional language is, according to the publication, both 'loved and encouraged' in the *Land*.[106] The Minister for Cultural Affairs holds that Low German is 'characteristic of the regional culture' in MV and that the protection of it is a central task for the government.[107] Thus, according to the MV constitution (translated into Low German in 1994), the *Land* should 'protect and support the preservation of the Low German Language'.[108] Low German is promoted through public support for competitions and contests for authors who write in Low German, for language competitions for children and adults, and for Low German theatre groups.[109]

---

102 StK MV (2002b), 'Zusammenarbeit des Landes mit Südschweden soll weiter ausgebaut werden' *Pressemitteilung Nr 69*, 2002-03-28, Schwerin; MfBWK MV (2001a:7).

103 MfBWK MV (2001a).

104 Low German is said to have been spoken more widely until the middle of the 20th century. As a written language, it has been regarded as more or less obsolete since the early 17th century. On 1 January 1999 the *Bundesrat* gave Low German the status as an official regional language in northern Germany. See LandtagsNachrichten MV (2004), 'Jubiläum, Vor zehn Jahren trat die Landesverfassung in Kraft', 1/2004, Schwerin; Rösler, Irmtraud (2004), 'Geschichte der Niderdeutsche Sprache', at *www.kulturportal-mv.de*, 2004-10-11. Institut für Niederdeutsche Sprache (1999), 'Plattdeutsch jetzt anerkannt als Regionalsprache', at *www.ins-bremen.de,* 2004-10-07.

105 Bahra (1997); Reich, Konrad (ed.) (2001), *Mecklenburg-Vorpommern: ein Porträt*, Rostock: Konrad Reich Verlag.

106 StK MV (2002c:31).

107 MfBWK MV (2002:6).

108 StK MV (2000d:art.16).

109 StK MV (2002c:31); Drs.3/2975 (2002-06-26), *Redebeitrag Minister Kauffold*, Schwerin.

The Need for Inner Orientation

Even if its location on the Baltic Sea, a rich cultural heritage, and the regional language are said to provide MV and its inhabitants with a firm regional identity, there are voices arguing that 'we' need to strengthen 'our' identity. We find the CDU group in the *Landtag* asking whether 'a campaign for strengthening regional identity' should not be initiated.[110] A promotion of the regional identity would, according to the CDU, bring 'mental orientation' and '*Heimatgefühl*'[111] to the inhabitants.[112] In a publication from the early 2000s, it is said that a 'solid regional identity', '*Heimatgefühl*' or 'a firm feeling of regional belonging' would lead to civic involvement and to people taking initiatives for the future development of the *Land*. This, it is argued, would contribute to an improvement of the future prospects of the *Land*.[113] In other contexts, it is suggested that an active cultural life and a strengthening of regional identity would counteract extensive migration flows from MV.[114] One CDU representative conceives culture as an 'extremely important identity-creating factor for the *Land*' as well as a 'central location factor'.[115]

Representatives of the SPD/PDS coalition government, too, emphasize the importance of regional identity formation. Preservation of distinctive regional features is thought to result in *Heimatgefühl,* regional identity and regional self-esteem.[116] Self-esteem is then promoted as essential, and as something that brings with it tolerance and openness.[117] The PDS Minister for Regional Development holds 'the feeling of togetherness and of joint responsibility' to be extremely important for regional development.[118] The SPD-led Ministry for

110 CDU Landtagsfraktion MV (2002:11).

111 Literally, 'a feeling of homeliness'.

112 CDU MV (2001); Landtag MV (2001c), Schnoor, Steffi in *Plenarprotokolle*, 3/63, 28 juni, Schwerin.

113 CDU MV (2001).

114 Drs.3/1212 (2000-03-29), *Antrag der Fraktion der CDU*, Schwerin.

115 Landtag MV (2002c), Schnoor, Steffi in *Plenarprotokoll*, 3/84, 26 Juni, Schwerin.

116 Landtag MV (2001b), Kauffold, Peter in *Plenarprotokolle*, 3/63, 28 Juni, Schwerin; MfBWK MV (2001b), 'Grußwort des Bildungsministers Prof. Peter Kauffold' *Pressemitteilung Nr 166-01,* Schwerin; StK MV (2002f); MfBWK MV (2003:2); Kulturforum Mecklenburg-Vorpommern (2002).

117 StK MV (2002f); Kulturbeirat MV (2002), *Erarbeitung einer Kulturentwicklungskonzeption für das Land Mecklenburg-Vorpommern*, Schwerin; Kulturbeirat MV (2003), *Grundsatzpapier zur Kulturentwicklung im Land Mecklenburg-Vorpommern*, Schwerin, p.5.

118 MfABL MV (2003b).

Cultural Affairs stresses, like the CDU above, that culture is a way to counteract the ongoing depopulation of the region. Support for the cultural sector is thus seen as a way to attract young people to the region, and as a means of making them stay and settle there.[119]

Calls for a strengthened regional identity and for preservation of the cultural heritage are to be found also at a subregional level.[120] In one subregion, a consultant was commissioned to draw up a marketing strategy for 'identity formation' in that particular subregion.[121] Hence, it was argued, 'identification with the region, its values and potentials is ... an important factor for overall economic development'.[122] In the other subregions, too, we find references to what an increased '*Wir-gefühl*'[123], 'cultural attachment' or 'identification with the *Heimatkultur*' would do for the subregion. *Heimatkultur*, cultural institutions and monuments are, together with the regional mass media, recurrently said to play a central role as transmitters of tradition and regional identity.[124]

Culture and Competitiveness

Just as in Norrbotten, culture and identity are said to be of importance for regional economic development. All parties represented in the *Landtag* have since the mid-1990s stressed the need to coordinate cultural policy, employment policy and economic policy. After a debate on these issues in the *Landtag*, the then CDU-led Ministry of Economics commissioned a study in which the cultural economic potential of the *Land* would be identified.[125] The report,

---

119 MfBWK MV (2003).

120 RPV Mittleres Mecklenburg/Rostock (2002:35).

121 Löhlein, B. und Löhlein, G. (2000) in RPV Meckleburgische Seenplatte (eds.), *REGIOnal*, No 1, Neubrandenburg.

122 Rettler, Bernhard (2000) in RPV Meckleburgische Seenplatte (eds.) *REGIOnal*, No 3, Neubrandenburg.

123 Literally, 'we-consciousness'.

124 RPV Vorpommern (1998:15, 154); Region Vorpommern e.V (n.d), *Vorpommern, Regionalmanagement, Regionalmarketing*, Stralsund; RPV Vorpommern (2001:3); RPV Mecklenburgische Seenplatte (1998), *Regionales Raumordnungsprogramm*, Neubrandenburg, p.127-8; Mecklenburgische Seenplatte (2000a:1). See also StK MV (1998:14).

125 When the question was raised for the first time (April 1995), the government was an SPD/CDU coalition. The Ministry of Economics, as well as the Ministry of Cultural Affairs, was then led by SPD representatives. In 1996, Ministry of Economics was switched from Harald Ringstorff (SPD) to Jürgen Seidel (CDU); see Werz, Niko-

delivered in 1997, concentrated on the direct effects of the cultural sector on the economy.[126] It concluded that cultural tourism could be further utilized, and that commercial use of old cultural traditions would be good for the cultural economy.[127] Yet the 1997 report is far from unique in its advice to go for cultural tourism. The PDS stresses in several contexts that cultural tourism is a means of developing the economy as well as of 'maintaining the cultural identity of the *Land*'.[128] The SPD-led Ministry for Cultural Affairs maintains that there is a general connection between culture and economy, and holds that 'different forms of cultural tourism are gaining increasing importance'.[129] Besides, several subregional actors stress the importance of an efficient use of 'cultural historic potentials' in the tourism sector.[130]

Regional specialities and delicacies are, just like cultural tourism, seen as an increasingly important line of business. Yet regional products are also understood as a means for marketing the region and as something that would benefit the development of a common regional identity. An SPD representative in Mecklenburg Lake District asserts that 'the more uniform the products on the global market turn out to be, the more important will it be to preserve one's own identity'.[131] A CDU representative in the *Landtag* follows the same line of argument, stressing that the development of products that are 'characteristic of the region' is an 'important step when aiming at a distinctive profile and identity for the region'.[132] Through the regional press, we learn that 'regional products are forming identities and are safeguarding job opportunities in the region'.[133] In Mecklenburg Lake District, however, the RPA maintains that the area lacks a

---

laus and Hennecke, Hans Jörg (eds.) (2000), *Parteien und Politik in Mecklenburg-Vorpommern*, München: Olzog Verlag, p. 274;

126 See also MfWi MV (1997) *Kulturwirtschaft in Mecklenburg-Vorpommern*, Schwerin.

127 MfWi MV (1997:IVff.).

128 Landtag MV (2001a); PDS MV (1997b), *Wahlprogramm der PDS MV zu den Landtagswahlen 1998*, Schwerin.

129 MfBWK MV (2002); MfBWK MV (2003:3).

130 RPV Mittleres Mecklenburg/Rostock (2002:35); RPV Vorpommern (1998:110,114); RPV Mecklenburgische Seenplatte (1998:127); Löhlein, Boris (2000), in RPV Meckleburgische Seenplatte (eds.) *REGIOnal*, No 3, Neubrandenburg, p.16; Rettler (2000).

131 Kautz, Michael (2000), in RPV Meckleburgische Seenplatte (eds.), *REGIOnal*, No 1, Neubrandenburg, p.1.

132 Drs.2/386.

133 OZ (2004a), 'Rostocker Senf passte nicht in Kühne-Konzept', 2004-02-06, p.5.

regional culinary heritage and that this is an inherent weakness of this particular subregion.[134]

Issues related to cultural economy have, apparently, influenced the political agenda ever since the debate first started in the mid-1990s. It is recurrently stressed that culture is a central 'soft location factor', and that culture and economy are increasingly interconnected.[135] In its election manifesto from 2002, the CDU regards culture as an important 'soft location factor'.[136] In the PDS programme from the same year we find calls for more explicit attempts to link culture and economy together.[137] During the same period, the SPD Minister for Cultural Affairs states that 'economic aspects gain more and more in importance' in cultural policy.[138] In their 2002 coalition treaty, the SPD/PDS coalition bring out culture as an important factor for economic growth and development.[139] The government, as well as representatives for the subregional level, refer to culture, self-esteem or 'a satisfaction with the *Land*' as 'essential qualities' or 'important pillars' if the region is to be able to cope with global competition.[140] We do find occasional protests against the increasing commercialization of the cultural sphere, but on a whole there is little opposition to this way of utilizing culture.[141] Also parties without representation in the *Landtag* stress the link between culture and economy. The Liberal Party (FDP) suggests in its election manifesto from 2002 that culture 'creates jobs' and that the party aims to make the *Land* 'fit for the future', partially through 'a better cultural marketing' of it.[142] The Green Party for its part accuses the ruling coalition of giving 'insufficient support to the

---

134 RPV Mecklenburgische Seenplatte (2000a:43).

135 Drs.2/319 (1995-04-18), *Kleine Anfrage der Abgeordneten Frau Schnoor*; Drs.2/386 (1995-05-12), *Antwort der Landesregierung*; Drs.2/525, 16.06.95, *Kleine Anfrage der Abgeordneten Frau Lautenschläger* (PDS); Drs.2/608, *Antwort der Landesregierung; Kultur und Wirtschaft*, 11.07.95; Landtag MV (2001a), Bartels, Gerhard in *Plenarprotokolle*, 3/63, 28 juni, Schwerin.

136 CDU MV (2002).

137 PDS MV (2002:4).

138 MfBWK MV (2002:10).

139 StK MV (2002e) *Vereinbarung SPD/PDS über die Bildung einer Koalitionsregierung für die 4. Legislaturperiode des Landtages von Mecklenburg-Vorpommern*, Schwerin, punkt 40,204,205; StK MV (2002f).

140 RPV Mecklenburgische Seenplatte (2000a:1); PDS MV (2002:2); MfBWK MV (2003:3). See also MfBWK MV (2001a); StK MV (2002g) 'Kultur baut Brücken im Ostseeraum', *Rede des Ministers President Ringstorff*, Schwerin.

141 Landtag MV (2002a), Schmidt, Karin in *Plenarprotokoll*, 3/84, 26 juni, Schwerin.

142 FDP MV (2002), *18 gute Gründe FDP zu wählen, Damit das Land nicht untergeht*, Schwerin.

cultural economy'. It claims that the cultural economy is an important interface between the economic, touristic and cultural development of the *Land*.[143]

Image and Image Campaigns

Just like in Norrbotten, we encounter a number of voices arguing that the image of the region is of central importance for its general development. In the 1997 report on cultural economy, it is maintained that a positive, 'cultural' image is 'a precondition for economic success for those working in the cultural economy'.[144] The SPD/PDS coalition government maintains in several contexts that the region's image is of great importance to attempts to attract investors to the region.[145] At a subregional level, too, it is stressed that, due to its significance for economic development, the image of the region should be a prioritized field of action.[146]

In several cases, a link is drawn between the external image of the *Land* and its internal identity.[147] It is not always clear whether the concept of 'identity' alludes to people's feelings of belonging or to the formal identity or character of the *Land* itself. The PDS Minister for Regional Development, however, stresses the link between a common regional identity and a 'distinct profile' that reflects the 'specific character of Mecklenburg–Western Pomerania'.[148] The CDU suggests, furthermore, that the formation of regional identities needs to be supported through a 'self-confident promotion of the *Land*'. Such promotion, it is argued, should influence the ways in which the *Land* is perceived by its own inhabitants as well as by outsiders.[149]

The general conviction that the image of a region is of vital importance for its overall development led to a number of image campaigns being launched in MV during the 1990s. The focus on images is indeed far from exceptional for the regions studied here. In Germany, we find image campaigns in as good as every

143 Grüne MV (2002a), *Rasender Stillstand: Kulturpolitik in MV*, 2002-02-04.
144 MfWi MV (1997:V).
145 Landtag MV (2001b); LandtagsNachrichten MV (2003a), 'Aus dem Plenum', 8/2003, Schwerin, p.5; MfABL MV (2003b).
146 RPV Mittleres Mecklenburg/Rostock (2002:35); RPV Mecklenburgische Seenplatte (1998:127); RPV Mecklenburgische Seenplatte (2000a:43); RPV Vorpommern (1998:34) Löhlein, B. (2000:16); Region Vorpommern e.V (n.d:1).
147 Löhlein, B. (2000:16); Region Vorpommern e.V (n.d:1), RPV Mittleres Mecklenburg/Rostock (2002).
148 Holter (2000b:16); MfABL MV (2003b).
149 CDU Landtagsfraktion MV (2002:37); CDU MV (2002).

Land. A short overview brings out image campaigns and slogans spanning from *Wir können alles ausser Hochdeutsch*[150] (Baden–Würtemberg) and *Laptop und Lederhosen* (Bavaria) to *Wir in NRW*[151] (North Rhine–Westphalia). In contrast to these regions, however, Mecklenburg–Western Pomerania was understood as being burdened by an extraordinarily poor reputation when the first regional image campaign was launched.[152]

Concern about the poor reputation of the *Land* came up in no uncertain terms in the image campaign launched by the MV government in 1997. One of the explicit aims of this campaign was to rectify the image of Mecklenburg–Western Pomerania as a backward place, dominated by unemployment and right-wing violence.[153] One part of the campaign was made up by a newspaper advertisement carrying a photograph of a young black woman. The banner of the advertisement stated '*Hier ist MV – Blond, Blauäugig, Blöd*'.[154] Underneath the photograph, the caption explained, first, that the girl in the photo was a student and born in the city of Greifswald (located in Western Pomerania), and second, that 'not everything is what it seems'. The campaign as such came to be heavily criticized by the Green Party in the *Bundestag* for portraying the woman in a sexist way, and by local citizens for feeding prejudices about people in Mecklenburg–Western Pomerania as being more stupid than others.[155]

After that campaign, the explicit goal of changing the negative image of the *Land* was buried for some years. Since 2000, however, several initiatives have been taken to promote the *Land* as a tourist region, characterized by health, recreation, pleasantness and joviality. One of the most widespread campaigns has been the *MV tut Gut*-campaign,[156] initiated in the early 2000s and aiming to establish the *Land* as 'a competitive "brand" in Germany and in the rest of

150 'We are capable of everything - except High German'.

151 'We in North Rhine–Westphalia'.

152 During 1995, the State Chancellery commissioned *Institut für Marktforschung* to undertake an analysis of MV's image in the rest of Germany. According to the analysis, MV was regarded as right-wing extremist, backward, and industrially and economically weak. This perception was primarily informed by the brutal pogrom against an asylum for refugees and guest workers in Rostock–Lichtenhagen in August 1992, and by some other incidents of right-wing violence. See Chapter 3. See also Bauer, Lars (2001), *Politische Kultur in Mecklenburg-Vorpommern*, Rostock: Institut für Politik- und Verwaltungswissenschaften, Universität Rostock.

153 StK MV (1997), 'Werbekampagne für Mecklenburg-Vorpommern', *Presseinformation Nr. 115/97*, Schwerin.

154 'This is MV – blond, blue-eyed, stupid.'

155 Bauer (2001).

156 'MV is good for you.'

Europe'. According to the government, the campaign should aim not only to promote the *Land* to those around. Rather, it was argued, 'it should contribute to strengthening the self-confidence of the *Land* itself' and to bringing together actors interested in promoting the region.[157]

Another campaign that aimed to project and image and to unify political and economical actors in the region was the one launched immediately after the announcement that Leipzig and Rostock would be nominated as the German candidate for the site of the Olympic Summer Games 2012.[158] Throughout the campaign, a number of public actors pointed out that the campaign should be of critical importance whether or not the IOC decided to support a further candidacy.[159] The campaign was widely thought to bring a higher level of public engagement and recognition, and to spread the message that this part of Germany is 'worthy to be reckoned with'.[160] The day after the IOC decision *not* to support the Leipzig/Rostock candidacy, the public transport companies in the region adopted the slogan 'Rostock has won', maintaining that the campaign had won for Rostock 'reputation, self-confidence and community feeling'. Other commentators held that the campaign had brought people 'closer to each other' and that it had 'proved that the whole region had been supporting the candidacy'.[161]

---

157 *www.mv-tut-gut.de*, 2004-10-07.

158 Rostock was part of the general Leipzig campaign, as the Olympic regatta would have taken place there if the candidature had been successful. Parts of the campaign were subsidized through the *Arbeitsmarkt- und Strukturentwicklungsprogramm* of the *Land*, and by financial guarantees from the *Land* government. A wide range of private actors combined their support for the candidacy with a promotion of their own business. Companies advertised in the regional press under slogans such as 'We support the campaign' or 'We back our region'. See Olymp-Club Rostock (2003a), *Bürger sollen Olympia-Bewerbung mitgestalten*, 2003-12-04, Olymp-Club Rostock (2004); *Land gibt Regierungsgarantien für Olympia* 11.12.2003; OZ (2004c), 'Various Advertisements', 2004-05-16.

159 OZ (2004b), 'Wirtschaft stärkt Olympia-Bewerburg', 2004-05-12; MfABL MV (2004b), 'Das Untermehmen bringt Alle weiter', *OstseeZeitung*, 2004-05-16.

160 MfABL MV (2004b); Pöker, Arno (2004), 'Gratulation an Mitbewerberstädte', *Pressemitteilung der Hansestadt Rostock*, 2004-05-19; OZ (2004d), 'Ringstorff: Wir gehen mit gutem Gefühl in diesen Tag', 2004-05-18; Vater, Ulrich (2003), 'An einem Strang ziehen', *NNN*, 2003-11-19.

161 OZ (2004h), 'Rostock hat gewonnen', 2004-05-19; OZ (2004g), 'Party-Sturm fegte Tränen weg', 2004-05-19; OZ (2004f), 'Olymp-Club macht weiter und setzt sich neues Vereinsziel', 2004-05-19; OZ (2004e), 'Aus der Traum', 2004-05-19.

Culture and identity are apparently thought to benefit society in a number of ways. Culture and identity are thought to bring *Heimatgefühl* and a positive and advantageous image of the region. Culture is understood as something that can be commodified and thus have a direct and favourable effect on the regional economy. Given these notions, we need, as in the Norrbottnic case, to ask what position new cultures or rival cultural identities have in the political debate.

Basically, 'immigrant' or 'non-regional' cultures are at no point in time blamed for being a problem of any kind. On the contrary, it is maintained that 'xenophobia, hostility towards foreigners, anti-Semitism and right-wing extremism' need to be counteracted.[162] The SPD/PDS government declares itself committed to work for 'promoting a dialogue between cultures', for the integration of immigrants and for improving immigrants' living conditions.[163] Children of 'non-German origin' should be helped with a better education, and immigrants and refugees should be cared for.[164] The then Minister of Cultural Affairs holds human rights to be vital, and stresses that 'a Europe of the regions requires respect for ... the emergence of cultural diversity within the regions'.[165] 'Dialogue among cultures ... acceptance, tolerance and mutual respect among different cultural identities' is, in other contexts, put forward as an important question for MV.[166]

Yet immigrant culture or alternative cultural identities are seldom referred to in discussions about the economic advantages a rich culture may bring. Indeed, the Minister for Cultural Affairs has mentioned that cultural influences brought to the *Land* by artists 'from other cultural spheres' is of general advantage for the development of the region, but in many cases multiculturalism, cultural diversity or migrant cultures are mentioned together with words such as tolerance, acceptance and the absence of violence.[167] A SPD representative in the *Landtag* stresses that an ongoing dialogue between the region's own culture and other, foreign cultures is a necessity in order to avoid conflicts and marginalization.[168]

162 Landtag MV (2003e), Voland, Angelika in *Plenarprotokolle*, 4/17, 26 juni, Schwerin.; MfBWK MV (2000), 'Kulturen im Dialog – gemeinsam gegen Gewalt', *Rede des Ministers Kauffold*, 2000-12-06, Schwerin.

163 StK MV (2002e:punkt 243).

164 StK MV (2002f).

165 MfBWK MV (2001a).

166 Kulturbeirat MV (2003).

167 MfBWK MV (2002).

168 Landtag MV (2003c), Lohse, Frank Ronald in *Plenarprotokolle*, 4/17, 26. juni, Schwerin.

The SPD-led Ministry of Cultural Affairs maintains that dialogue between cultures is the path to peace and non-violence.[169] Pluralism, solidarity and internationalism are important keywords when the Minister of Cultural Affairs presents a project named 'Cultures in Dialogue – together against violence'.[170] In yet another context, the minister points out how important 'peaceful coexistence among different nationalities' is.[171] The PDS, too, tends to refer to multiculturalism particularly in discussions about anti-racism and human rights.[172]

Culture, Identity, Image and Pride

According to Scott, there is an evident and deepening tension between the understanding of culture as something 'narrowly place-bound' and as 'a pattern of non-place globalised events and experiences'.[173] To judge from the documents analysed here, culture is understood as being *both* place-bound *and* mobile. In some cases, culture is described as a process in constant flux. In both Norrbotten and Mecklenburg–Western Pomerania, we find references to what a dynamic cultural life would do for the creative power in the region. And in both regions, culture is represented as an increasingly important line of business. In these cases, a rather unemotional and more inclusive understanding of culture is involved.

Still, regional culture and identity is just as much 'romanticized [and] stereotyped'.[174] References to history and folklore, 'roots' and cultural distinctiveness do permeate the policy discourse in both regions. Particularly in Norrbotten, we find a variety of references to genuine, inherited and place-specific regional culture. We find references to the distinctive 'character' of the Norrbottnic population and to a lifestyle that is said to dominate this particular region. As in early romantic nationalism, there seems to be a conviction that culture, lifestyle, and social institutions are expressions of 'the "soul", the

169 Kulturbeirat MV (2002), *Erarbeitung einer Kulturentwicklungskonzeption für das Land Mecklenburg-Vorpommern*, Schwerin.

170 MfBWK MV (2000).

171 MfBWK MV (2002).

172 PDS MV (1997a), *Parteiprogramm des PDS 1998*, Schwerin.

173 Scott, Allen J. (2000), *The Cultural Economy of Cities*, London: SAGE, p.3.

174 Keating (1998:84).

"spirit", or the "character" of a people'.[175] We do also find a wealth of references to the alleged effect of nature and climate on the regional genius. This, it has been claimed, is also an argument that 'figures strongly in romantic nationalism'.[176]

The discourse on culture in Mecklenburg–Western Pomerania is markedly less romantic than that in Norrbotten. We do find references to a rich cultural heritage and to the need for regional belonging, but the way in which culture is delineated here reflects pragmatism rather than romanticism. Narratives about regional genius or the regional way of life do crop up in MV, particularly in references to the Low German tradition. Still, much more attention is paid to the alleged need for a profitable cultural economy. In both regions, however, the development of a cultural economy is thought to be closely linked to the preservation of a place-specific cultural heritage. Cultural heritage is indeed thought to impart identity to the inhabitants, but is also thought of as a commodity to be marketed to tourists and potential settlers.

Cultural Hierarchies?

The linking of culture to economy is found in EU cultural policy, in state level policy, and in a variety of European regions. Moreover, the assumption that there is a link between culture and economic activity has achieved wide support also in regional economics.[177] In some cases, a rich cultural life is understood to function as a potential magnet attracting skilled workers or investors to the region.[178] Culture and cultural industries are ascribed the capacity to provide regions with identity and profile, or to create 'self-confidence in communities, which can improve prospects for endogenous growth'.[179] These notions are

175 Tollebeek, Jo (1998), 'Historical Representation and the Nation-State in Romantic Belgium (1830-1850)', *Journal of the History of Ideas* 59(2), pp.329-353, p.330; Guibernau (1999:23).

176 Hamilton, P. (2002), 'The Greening of Nationalism: Nationalising Nature in Europe', *Environmental Politics* 11(2), pp.27-48.

177 Mayo (2000:111ff); Power (2002:103); Keating *et al.* (2003).

178 Wikhall, Maria (2003), 'Kultur och regional attraktivitet – migrationsbeslut bland högskoleutbildade i Sverige', in Egeland, H. and Johannisson, J. (eds.), *Kultur Plats Identitet*, Nora: Nya Doxa

179 Bryan, J, Hill, S, Munday, M, Roberts, A. (2000), 'Assessing the Role of the Arts and Cultural Industries in a Local Economy', *Environment and Planning A*, 32, pp. 1391-1408, p.1391. See also Lindeborg, Lisbeth (1995), *Regionalt Samarbete i Europa – med tyska erfarenheter*, Stockholm: Fritzes.

linked to the idea that it is important for regions to be recognized as a 'sticky' or even 'magnetic' place. According to Markussen, stickiness connotes the ability to attract investors, workers and industries to a certain location and to keep them there.[180] In much the same manner, magnetism has been used to describe places or regions 'full of action, to which people are drawn, where it is beautiful [and] where culture blossoms'.[181] The role of culture in these 'sticky' or 'magnetic' places is thus primarily of an *indirect* kind.

Yet the *direct* economic significance of cultural or heritage industry has also been considered substantial. In politics as well as in research, we meet with examples of how the 'genuine', the 'authentic', the 'regional' or 'distinctive regional characters' are attributed great importance for regional economic development.[182] In many peripheral regions – including those studied here – it is greatly hoped that cultural tourism will get the regional economy going. We also find a wealth of examples of how regional actors try to promote or commodify an alleged regional cuisine for the very same reasons.[183]

Certainly, one could ask whether such things as 'authentic' or 'genuine' cultures really exist.[184] This question is even more warranted when one talks about cultures that are said to be linked to a certain territory. The most convincing position in this matter, I would say, is that culture is a changeable social construct. As such, it is being both reconstructed and adapted to situations encountered. The changing and receptive character of culture subsequently turns the idea of 'pure' or 'preserved' regional cultures into a contradiction in terms. The marketing of genuine or authentic regional culture could thus be criticized as a self-contradiction. Hence, if we believe in the existence of authentic cultures and then let PR agencies promote and commodify these cultures, they simply cease to be authentic in the full sense of the word.

Yet the main concern here is not to decide whether or not the cultural traits referred to by regional policy-makers really can be called 'authentic'. Much more important is the fact that the desire to commodify 'typical' regional cultural characteristics implies that different cultural expressions are handled in different

180 Markussen, Ann (1996), 'Sticky Places in Slippery Space: A Typology of Industrial Districts', *Economic Geography* 72(3), pp.293-313.

181 Nilsson, Marianne and Brulin, Göran (1997), *Identiprenörskap - företagande med regionalt ursprung*, R 1997:22, Stockholm: NUTEK.

182 Nilsson and Brulin (1997:3,54).

183 Salomonsson, Anders (1994), 'Piteplat, Värmlandskorv och Äggakaka', in Blomberg, B. and Lindquist, S.O (eds.), *Den Regionala Särarten*, Lund: Studentlitteratur, p.233.

184 Hylland-Eriksen (1999).

ways. We find references to culture as something 'narrowly place-bound' side by side with references to culture as mobile 'events and experiences'.[185] Nevertheless, there is an implicit evaluation of these two types of culture. In many regions, efforts made to 'exploit a real or imagined past as a tourist attraction'[186] have led to a 'celebration of "difference"'.[187] As the particular, the inherited and the genuine are to such a great extent seen as assets in a growing cultural industry, these peculiarities turn out to be far more precious than immigrant, non-regional cultural expressions.

The regions are indeed described as 'diverse' and 'multicultural'. Nonetheless, we do need to ask how this diversity is defined and whether 'one diversified manifestation [is] more deviant than another?'.[188] In Norrbotten, it is clear that the much-lionized diversity is thought to consist of three supposedly distinct cultures. The mix of Saami, Swedish and Finnish cultures is represented as an exotic comparative advantage for the region. And in Mecklenburg–Western Pomerania, it is primarily various forms of cultural heritage that are thought of as a resource in the cultural economy. The immigrant 'multiculture' is indeed accepted and tolerated, but is to a far less extent promoted as a resource for the development of the region. Accordingly, the commodification of culture implies not only that stereotyped images and notions of regional culture are being created and distributed.[189] The commodification of 'a real or imagined' regional culture and past involves also a stratification of different cultural expressions.

Identity and Identity Politics

What we can observe in both Norrbotten and MV is a rather contradictory approach to regional identities. Regional identities are on the one hand referred to as something inherited. Regional identities are then understood as being path-dependent, formed by cultural roots, by the landscape, and by (parts of) the

185 Scott (2000:3).

186 Keating (1998:84).

187 Bond, R., McCrone, D., Brown, A. (2003), 'National Identity and Economic Development: Reiteration, Recapture, Reinterpretation and Repudiation', *Nations and Nationalism* 9(3), pp.371-391, p.373.

188 Hellström, Anders (2003), '(Re)making European Unity', in Petersson, Bo and Clark, Eric (eds.), *Identity Dynamics and the Construction of Boundaries*, Lund: Nordic Academic Press, p. 182,196.

189 Löfgren (1989:11); Löfgren, Orvar (1994), 'Kurhotell, Pensionat och Sommarland' in Blomberg, B. and Lindquist, S.O (eds.), *Den Regionala Särarten*, Lund: Studentlitteratur, pp.140-151.

regional history. On the other hand, there is an ambition to strengthen these inherited and path-dependent identities by means of campaigns, information, and support for regional culture. So the approach to identities found here is actually a mix of essentialist path-dependency and over-optimistic constructionism.

The attempts to strengthen regional identity could easily be understood as instances of identity politics.[190] In social theory, identity politics have commonly been understood as the politicization of groups that strive for emancipation.[191] From that perspective, identity politics can be understood as an attempt to release a group of people from oppression. Identity politics can, however, also be interpreted as an attempt to win loyalty for existing or newly established power structures. Nation building or the politics of identity pursued by the European Union has oftenly been understood in the latter way.[192] Yet there are situations in which it is not completely clear whether an identity political project is steered from above, or whether it derives from a broader part of a certain population.

Here, we need to ask how identity politics at the regional level is to be understood. Is regional identity politics primarily an emancipatory act, or is it rather an attempt to legitimize a political project, similar to some earlier nation-building projects or to contemporary attempts to strengthen the feelings of belonging to the European Union? Indeed, the answer to that question depends on whether or not we understand the region as an oppressed community of common interests. If we understand the region and regional identities as primordial categories faced with oppression, the identity politics pursued by regional representatives may be interpreted as a means of emancipation. If, as an alternative, we understand regions as political (elite) constructs and acknowledge the multitude of identities and interests found within these regions, identity politics is rather to be understood as an attempt to legitimize a political project.

Many previous scholars have emphasized the vagueness of the concept of identity.[193] Yet it has become widely accepted that identities are situational, relative, flexible and changing, but that they nonetheless tend to require some

---

190 Süssner, Josefina (2000), 'Regionen som identitetspolitiskt projekt', *Nordisk Samhällsgeografisk Tidskrift*, Nr 31.

191 Bondi, Liz (1996), 'Locating Identity Politics', in Keith, M. and Pile, S. (eds.), *Place and the Politics of Identity*, London and New York: Routledge; Hansen, Peo (2000), *Europeans Only? Essays on Identity Politics and the European Union*. Umeå University, p.51; Mayo (2000:46ff).

192 Hansen (2000).

193 Billig (1995:8); Keating (1998:85-87); Paasi, Anssi (2003a), 'Region and Place: Regional Identity in Question', *Progress in Human Geography* 27(4), pp.475-485, p.480.

sort of categorization.[194] The idea that individuals are capable of carrying many – overlapping, complementary or even seemingly contradicting – identities has also achieved the status of conventional wisdom.[195] When one analyses identity politics rather than identities, however, some of this complexity disappears. Hence, whereas seemingly contradictory identities can coexist in the same individual or the same group, the whole point of identity politics is to strengthen one particular identity and to reinforce one particular category at the cost of some others. The political ambition to strengthen *regional* identity is an ambition to 'make' the regional group and to 'unmake' other groups, or at least to submit other feelings of belonging and solidarity to a less central object.[196] In that respect, the ambition to support regional identities must be understood as an attempt to get support not only for a particular territory but also for a political project linked to that particular territory.

Every construction of a collective 'we' is based on the inclusion of those who fulfil the requirements of the group and the exclusion of those who do not. Accordingly, the construction of identities is intrinsically intertwined with the construction of boundaries.[197] As identity politics involves the definition of boundaries, groups and categories, we ought to ask what the primary requisites for 'membership' in these groups are.[198] Even if a regional identity is a political and social construct, the content that political actors ascribe to it is not open-ended.[199] The requirements for belonging to a *national* community have repeatedly been criticized for being both narrow and exclusive.[200] Yet a regional identity might very well be even more excluding and selective. When asking who may belong to Norrbotten or Mecklenburg–Western Pomerania, we will find out that it is an arbitrary category. Theoretically, 'the Norrbothnians' could be every resident in the County. Yet the cultural romanticism in the regional rhetoric thwarts that. The stereotypical image, particularly of the Norrbothnian, excludes those who fail to have the right kind of origin or the right kind of mentality. In that sense, the strategy of achieving regional development through regional identity politics entails some rarely examined implications of exclusion.

---

194 Lecours (2001:60); Hylland-Eriksen (1999:31); Keating (1998:87).

195 Paasi (2003a:475).

196 Bourdieu (1991:221).

197 Petersson and Clark (2003:12).

198 Bourdieu (1991:221); Guibernau (1999:14); Smith, A.D (1990), 'The Suppression of Nationalism?', *International Journal of Comparative Sociology*, XXXI, 1-2, p.8.

199 Smith (1990:8).

200 Hall, Stuart (1993), 'Rasism som ideologisk diskurs', *Häften för kritiska studier,* 3, Stockholm, p.10.

8. The Target Group of Regionalism

Throughout history there have been a number of social and political movements, designed to represent new or previously neglected groups in society. Among them we find global movements with far-reaching demands and achievements along with local initiatives with limited claims and results. So far, there are no indications that the regionalism studied in this book is based on, or even part of, any broader political or social movement.[1] Still, we should ask what groups, collectives or social categories regionalist ideology seems to concentrate on. That is, we need to investigate to *whom* its messages are addressed.

In the definition of regionalism adopted here, the regional *population* was fundamental.[2] Tentatively, we could then presume that the regional population is the actual target group of regionalism. Still, we need to explore how this 'population' is defined in each case. Thus, research on nationalism demonstrate that this category often turns out to be rather excluding in practice.[3]

In Chapter 7 we saw examples of a rather romantic way of understanding regional culture and a somewhat primordial approach to regional identities. Multiculture or cultural influences from outside the region are certainly not thought to pose threats to the region. Yet it is inherited cultures and identities that are brought out as being the most crucial. Given that, we may presume that the target group for the regionalism found here is constituted by those who can identify themselves with the somewhat essentialized representations of 'Norrbothnians' or 'Mecklenburg–Western Pomeranians'. Still, the fact that economic growth is so clearly accentuated in the regional policy discourse may indicate that those who are thought of as capable of contributing to economic growth and development will get extra attention in the policy discourse too.

1 See Mayo (2000:49).
2 Williams (1997:114).
3 Billig (1995); Painter (2000).

In Chapter 7, we came across several references to an alleged Norrbottnic culture and identity. Some of the voices raised there claimed that people in Norrbotten share positive characteristics such as straightforwardness, generosity, reliability and loyalty. That particular representation of the regional citizen is, however, far from consistent. In more than one context, we come across the idea that people in Norrbotten are particularly 'whining' or passive, and that there is a lack of civic spirit in the region.[4] Passages in the regional press tell us that Norrbottnic municipalities 'lack an entrepreneurial spirit', that the 'envy is awfully deep', and that 'Norrbothnians in general and people from the inland in particular belong to a terribly whining species', or tend to 'moan over the negative migration flows'.[5] Besides, Norrbottnic social life has been thought of as guided by an old-fashioned, male chauvinist attitude that makes young women leave the county in great numbers.[6]

Character Flaws or Stereotypes?

In parts, the assertion that people from the region are introverted, reserved or passive is dismissed as a deceptive stereotype imputed to the region by outsiders.[7] These images are nevertheless thought to be a burden for the region. The image of being a region marked by 'dole dependency, strikes ... melancholy, poaching, illicit distilling of alcohol' and harsh climatic conditions is, according to the Social Democratic Party, a problem for the entire region.[8] A local Social Democrat asserts that some tend to depict people from the region as 'whining, discontented "country folk"[9]'.[10] The chief editor of one of the regional dailies claims that the region is burdened by an image of being a place where 'people just grumble and complain, live on social allowances and demand that other

4 NSD (2001d), 'Arbetskraft till 15 000 jobb sökes', 2001-11-23; for a critique of that, see Sundberg, Robert (2001), 'Ökad regional egoism missgynnar Norrbotten', *NSD*, 2001-11-24.

5 NSD (2001e), 'Fattiga S-kusiner från landet mötes på kongress', 2001-11-26; NSD (2001a), 'Företagsamheten spirar i Malmfälten', 2001-09-10.

6 NSD (2001c), 'Segelström varnar för Törnmans machokultur', 2001-11-09.

7 Framtidslandet (2004).

8 SAP Norrbotten (1997:3).

9 *Glesbygdsbor* in the original.

10 Forsberg, R. (2004), 'Framtiden finns i Kalix', *NSD*, 2004-09-23.

people take care of the development of the region'.[11] A County Governor calls into question 'the myth about Norrbothnians as quiet and reserved' and retorts that people from the region are 'open, generous and very straightforward in their manner'.[12] When the character flaws are conceived as mere stereotypes, the policy recommendation turns out to be a spreading of a 'true' and more positive image of the region.[13] Thus, as indicated in Chapter 7, several initiatives have been taken in order to correct the allegedly biased and negative image of Norrbotten.

Still, it is recurrently stressed that there *is* a certain mentality in Norrbotten that is to the disadvantage of the region when it comes to economic development. One of the alleged drawbacks in the region – understood as extremely important to overcome if the regional economy is about to develop – is the so-called *Jantelag*.[14] In *The Guide*, we find references to the injurious effects that the *Jantelag* has had on the region and, to judge from quotes in the regional press, it is not unusual for local politicians to blame the people for letting the *Jantelag* rule the area.[15] The *Jantelag* has been cited to explain why entrepreneurs from within the region decide to move and to settle in other regions; and an editorial column in the regional daily contends that the *Jantelag* circumscribes patterns of behaviour both in the public administration and among ordinary people.[16] In another column, the same editor holds that there is a 'we-can't-be-bothered attitude' in Norrbotten, said to 'complement the *Jantelag* and the enviousness in the region in a rather disadvantageous way'.[17] The regional

11 Gröning-Degerlund, L. (2001c), 'Suck och stön ska sätta Norrbotten på kartan', *NSD*, 2001-10-05.

12 Eriksson, P.O (2003a:4).

13 NSD (2003b).

14 The law of Jante (*Jantelagen* in the original) is an idiom originally found in the novel *A Refugee Crosses his Tracks* by the Norwegian/Danish author and novelist Aksel Sandemose. There, 'Jante' is described as a town ruled by an informal but still very oppressive law that prevents the inhabitants from showing their own superiority to others. The law dictates, in short, that '(y)ou shall not believe that you are something, that you are as good as we, that you are wiser or better than us, that you are greater or know more than us, or that you can teach us anything at all'. Sandemose, Aksel (1933), *En Flyktning Krysser Sitt Spor*, Oslo: Tiden Norsk Førlag.

15 Partnerskap för Norrbotten (1995:14); NSD (2001e); Ohlsson, Staffan (1998b), 'Jantelagen', *Hjärtslag*, Nr. 4, p.8; Holmqvist, Per (1998b), 'Flitiga företagare i Övertorneå' *Hjärtslag*, Specialnummer, p.7.

16 NSD (2003c), 'Jantelagen styr Norrbotten', 2003-08-13; Gröning-Degerlund, L. (2003c), 'Några arbetstips, P.O Eriksson', *NSD* 2003-08-05.

17 Gröning-Degerlund, L. (2003d), ''Vi ids inte!' är fel inställning', *NSD* 2003-08-06.

population is, furthermore, understood to suffer from a poor attitude to service. Therefore, it is suggested, many of those working in service industries need to 'attend a course at Charm School'.[18] When the situation in one of the most problem-ridden municipalities in the region is debated in an article signed by a local politician, the editor replies by asking whether the complaining article is motivated by self-pity or by the hope of being pitied by others.[19] After being criticized for these statements, the editor insists that the *Jantelag* tradition, the moaning, the poor attitude to service and the poor entrepreneurial climate in the region can be substantiated by facts and figures.[20] In other contexts, however, the *Jantelag* is not seen as a problem for Norrbotten only, but as something that afflicts the rest of Sweden too. Another editorial writer stresses that the *Jantelag* 'hampers and prevents Sweden and Norrbotten from growing, as it makes it hard for single individuals to stand up for good ideas that would contribute to further development'.[21]

Apart from the *Jantelag*'s alleged ascendancy in the Norrbottnic mentality, we learn that 'self-sufficiency and quarrelsomeness' are problematic characteristics of the people in the region.[22] Leading politicians from the region are reported to regard 'the whining and grumbling for grants and subsidies', 'the lack of self-confidence' and the propensity to 'expect others to solve our problems for us' as central problems.[23] A representative of the Left Party states that it is time to 'leave the carping criticism behind, and to bring the meaningless negativism and the primitive complaints to an end'.[24] A representative of the Liberal Party stresses in much the same way that it is time to 'stop moaning and to stop bottling up the anger'. Thus, it is contended, a change of attitude is needed if Norrbotten is going to develop.[25]

In cases where the Norrbottnic mentality is understood as a genuine problem and not as one imagined in biased preconceptions, it is occasionally explained by reference to historical circumstances.[26] In *The Guide*, it is rhetorically asked why there is 'a spirit of enterprise' in some parts of Sweden but not in Norrbotten,

---

18 Gröning-Degerlund (2003d).

19 Gröning-Degerlund (2003b).

20 Rönnberg, Sten (2003), 'Journalister bör hålla sig goda för att fiska efter popularitet', *NSD*, 2003-08-11, (reply from Gröning-Degerlund under the same headline).

21 Wennberg, Jenny (2003), 'Avskaffa Jantelagen!', *NSD*, 2003-06-23.

22 Framtidslandet (2004).

23 NSD (2002a).

24 Lööw (2002).

25 Grönlund, Anna (2003b), 'Norrbotten kan växa och utvecklas', *NSD*, 2003-06-02.

26 Norrbottengruppen (1999:22).

and the tentative answer is that it 'has to do with ancient social and cultural traditions'.[27] Other attempts to explain the mentality in the region refer back to the historical influence of large-scale enterprises in the region.[28] The Centre Party concludes in this context that the dominance of large-scale enterprises in the region has combined with a 'wage-earner culture'[29] to keep out 'a developed culture of entrepreneurship'. This tradition, it is argued, is an obstacle to restructuring the regional labour market.[30]

A New Attitude

When the regional 'mentality' or 'mindset' is considered a substantial obstacle to economic development, it is not primarily image campaigns that are called for. Instead, what is demanded is a real change of attitudes within the region. Positive and constructive thinking and a 'capacity to focus on opportunities rather than on problems' are then brought out as preconditions for regional success.[31] Many voices stress that the regional population must become more interested in education and entrepreneurship, not least because the region is thought to be entering a post-industrial, knowledge-based era (see Chapter 4).[32] In *The Guide*, it is emphasized that (we) need to start processes 'capable of strengthening our faith in ourselves and the county'. By means of 'educational efforts, cultural achievements and seminars', it is argued, 'we shall link ourselves together as a people in a more explicit and purposeful manner'. That, it is contended, will be needed for 'meeting the great challenges of tomorrow'.[33] According to the County Administrative Board, the entire region needs to be more competent and qualified than it is at present. One reason for increasing the standard of attainment in the region is that 'competence attracts competence' and that 'smart individuals' would be more willing to move to Norrbotten if the know-how in the region expanded.[34] Also in other contexts it is argued that individuals' attitudes, competence, level

27 Partnerskap för Norrbotten (1995:30).
28 Framtidslandet (2004).
29 *Löntagarkultur* in the original.
30 Centerpartiet Norrbotten (1998:6).
31 Hultin, Bo (2003), 'Se möjligheter, inte svårigheter', *NSD*, 2003-01-02.
32 SAP Norrbotten (1997:3); Lindström (2001).
33 Partnerskap för Norrbotten (1995:14).
34 Lst Norrbotten (2004b:3).

of education and entrepreneurial spirit are important conditions for economic development.[35]

The well-known argument that regions gain from strong networks between small enterprises is found also in Norrbotten. In *The Guide*, we learn that small enterprises that 'derive their livelihood from local conditions and networks where people know each other' will constitute the future of the Norrbottnic economy.[36] The region is also said to need a climate of creativity and entrepreneurship in order to succeed. Thus, it is argued, the economy in a region is highly dependent on civic spirit.[37] In the 2000-2002 RDA, economic growth is understood to depend on intellectual and social capital. Social capital is then defined in terms of 'entrepreneurship, ability to cooperate, creativity and willingness to change'. The region needs 'an infrastructure which is promoting dynamism – not only in a technical but also in an attitudinal sense'. Yet, it is argued, 'cultural and social structures are far more difficult to change than economic ones'.[38]

Entrepreneurial Spirit

Opinions seem to divide over the historical position of private enterprise in Norrbotten. The Social Democratic Party holds that small enterprises constitute the backbone of the economy in the inland parts of Norrbotten, and the County Council stresses that 'in our county, we have entrepreneurs, will and creativity'.[39] In *The Guide*, however, we learn that the average Norrbothnian has traditionally been rather reluctant to engage in private enterprise. This deficiency is however thought to have improved during the last ten years and, according to *The Guide*, this 'positive attitude' is to be 'promoted and further developed'.[40]

It seems to be widely assumed that entrepreneurship will be even more crucial in the future and that the region must in any case strengthen its entrepreneurial spirit.[41] A report from the County Council and the Regional Association of Local

---

35 Centerpartiet Norrbotten (1998:10); Wikman (2003).

36 Partnerskap för Norrbotten (1995:10).

37 Partnerskap för Norrbotten (1995:12); Regionutredningen Norrbotten (1999b:10), (2000a:9).

38 Norrbottengruppen (1999:21, 23, 25).

39 SAP Norrbotten (2001:10); Lst Norrbotten (2004b:3).

40 In *The Guide* however, it is stressed that the city of Piteå has a tradition of private enterprising. Partnerskap för Norrbotten (1995:30).

41 Partnerskap för Norrbotten (1995:22); Folkpartiet Liberalerna Norrbotten (2002b); Wikman (2003); Lindström (2001); Holmqvist (1998a:3).

Authorities states that development cannot be created by administrative initiatives but is a result of individuals' ideas, values, and attitudes towards development, knowledge and new technology.[42] In the 2000-2002 RDA, measures capable of increasing 'the Norrbothnians' eagerness for education and entrepreneurship' are highly recommended.[43] The Social Democratic Party states in its party programme from 1997 that an image of Norrbotten as a successful region presupposes an 'open attitude towards entrepreneurs and private enterprises'.[44] Similarly, the Centre Party states in its 1998-2002 party programme that a new and 'enterprise-friendly spirit' needs to be created.[45] The Liberal Party stresses that it aims at creating 'civic engagement and an entrepreneurial spirit' rather than supporting an 'ever-increasing public sector'.[46] The spirit of enterprise 'cannot be forced', it is maintained, but politicians still need to 'back up energetic and motivated entrepreneurs, and encourage them to realise their ideas'.[47] In this matter, a representative of the Liberal Party holds that the region needs a better entrepreneurial climate and that local and regional politicians are in a position to encourage it.[48] A Moderate Party representative concurs that the *Jantelag* 'needs to be abolished' and that entrepreneurs 'need to be given more space in our Sweden'. The lack of entrepreneurialism is understood as a problem not only for Norrbotten, but for the whole of Sweden.[49]

The Role Model

In some cases, we find regional actors emphasizing the need for role models or good examples, in politics as well as in private business. In *The Guide* the local entrepreneur is understood as one of the actors who will bring economic success to the region. The success of the local entrepreneur is, however, understood to be dependent on other people in the region changing their attitude towards private enterprise.[50] In the 2000-2006 Single Programme Document, an 'insufficient number of entrepreneurs and role models', as well as 'the small number of

42 Regionutredningen Norrbotten (2000b:2).
43 Norrbottengruppen (1999:33).
44 SAP Norrbotten (1997:3).
45 Centerpartiet Norrbotten (1998:6).
46 Folkpartiet Liberalerna Norrbotten (2002b).
47 Folkpartiet Liberalerna Norrbotten (2002a:5)
48 Grönlund, Anna (2003a), 'Ryck upp er, politiker', *NSD*, 2003-05-14.
49 Wikström, Åsa (2003), 'Vart tog de vackra orden vägen?', *NSD*, 2003-10-15.
50 Partnerskap för Norrbotten (1995:10,22).

potential leaders, entrepreneurs and other key figures', are referred to as threats and weaknesses for the northernmost parts of Sweden.[51]

Many voices maintain that attitudes towards entrepreneurs need to be reconsidered and that small-scale business enterprises should be treated more generously.[52] The Centre Party stresses that young peoples' 'enterprising spirit' must be encouraged rather than suffocated at school.[53] A Centre Party representative maintains that Norrbotten must adopt 'a more positive attitude towards private entrepreneurs', that private enterprise and entrepreneurs should be 'greeted with curiosity and enthusiasm' and that it is important to 'encourage young people to set up their own business'.[54] A year later, the very same representative reiterates that a 'change of attitudes towards present and future entrepreneurs' would be of the utmost importance for the development of the region. Hence, it is argued, we need to 'stimulate qualities such as an enterprising spirit ... and by all means counteract the *Jantelag*'.[55] The Liberal Party makes an equally strong point of encouraging entrepreneurship and commercial initiatives early on in life.[56] In its 1998-2002 party programme, the Liberal Party states that the Norrbothnian ought to 'view small businesspeople as heroes and to hail them as such'.[57] The Social Democratic Party points out in its 1997 party programme that 'small businesspeople' should be conceived not as 'potential big capitalists or tax evaders' but as 'hard working people ... who are central to the development of the region'.[58]

Not only entrepreneurs but also politicians are thought of as potential 'good, regional examples'. It happens indeed that politicians accuse each other of whining too much and of doing nothing but complain.[59] In this matter, a leading Liberal maintains that Norrbottnic policy-makers should 'get together and start looking for opportunities instead of only focusing on the obstacles'. Politicians should 'radiate a belief in the future' rather than complaining about the situation in the region.[60] A former County Governor and Social Democratic government minister contends in this respect that 'politicians who promote opportunities

---

51 Lst Norrbotten *et al.* (2000:33).

52 Zakrisson (2003); Lindström (2001); Folkpartiet Liberalerna Norrbotten (2002a:5).

53 Centerpartiet Norrbotten och Västerbotten (2001).

54 Nilsson, Frank (2002), 'Bra välfärd börjar i småföretagen', *NSD*, 2002-05-28.

55 Nilsson, Frank (2003), 'Vitalisera välfärden', *NSD*, 2003-10-10.

56 Folkpartiet Liberalerna Norrbotten (2002b).

57 Folkpartiet Liberalerna Norrbotten (1998:4).

58 SAP Norrbotten (1997:3).

59 Petersen, Karl (2001b), 'Sluta gnälla, Lars Törnman', *NSD* 2001-12-18; Grönlund (2002); See also Gröning-Degerlund (2001b); Gröning-Degerlund (2001c).

60 Grönlund (2003a).

instead of always pointing out problems' are among the prerequisites for economic growth and development.[61] Instead of parties and politicians 'moaning and groaning', it is argued in other contexts, Norrbotten needs politicians who are capable of 'seeing the prospects rather than the problems' and willing to 'advertise Norrbotten as a county of the future'.[62]

The Norrbottnic Diaspora

Apart from entrepreneurs and role models, there is yet another group that is given a topical interest in the policy discourse, namely, the (potential) home-comers. Since great parts of Norrbotten have been affected by negative migration flows during recent decades, and since the shrinking population is an overriding problem for the region, this is perhaps no surprise. Much policy-making in Norrbotten does in fact aim to end the negative migration flows, and several purposeful attempts have been made to entice emigrant Norrbothnians back to the region. In this matter, the Social Democratic Party stresses that it must be possible to combine the 'youthful desire to further self-fulfilment and development elsewhere' with a willingness to return to Norrbotten later in life.[63]

The 'Norrbotten Days', referred to in Chapter 7 as part of an initiative aiming to present a positive image of Norrbotten, is an outstanding example of how Norrbottnic political actors try to entice former Norrbottnic residents back to 'their' county. Even if it is emphasized that 'all others' are welcome, it is evident that the main target group of these events is potential home-comers, people who have left Norrbotten for better prospects in the southern parts of Sweden.[64] 'Emigrant Norrbothnians' are mentioned as a particular target group by the initiators, and a report from one of the participating municipalities claims that the event provides 'former residents of the municipality' with the opportunity to make contact with their former hometown.[65] 'With the Norrbotten Days', the organizers declared, (we) want to 'exhibit ourselves', to 'prove that we exist',

---

61 Rosengren, Björn (2003), 'Se på möjligheterna', *NSD Affärsliv*, 2003-10-17.

62 Gröning-Degerlund (2001c); Öberg, Maria (2001), 'Står kirunapartiet och Törnman för utveckling?', *NSD* 2001-10-16.

63 SAP Norrbotten (2001:6); Rosengren, Björn and Zakrisson, Kristina (2001), 'Norrbotten – Framtidslandet', *PT*, 2001-04-06; Lindgren and Riklund (2004); Grönlund (2003a).

64 Norrbotten Presenterar (2003), *Norrbottensdagen på Gärdet - Exponeringserbjudande*, Luleå.

65 Lindgren and Riklund (2004); Övertorneå Kommun (2002).

and 'show emigrant Norrbothnians that the home municipality and the home county is an attractive place to live'. The aspiration was to make more people return to Norrbotten, even if, as it is concluded, 'we welcome all others as well'.[66]

During the events in Stockholm, participants in the Norrbottnic cultural project 'Land of the Future' arranged interviews with so-called Norrbothnians in exile. These 'emigrant' Norrbothnians were equipped with cameras and tape-recorders and then encouraged to ask 'other Norrbothnians in exile' about 'how they perceive Norrbotten today, about homesickness and about how Norrbothnians are looked upon in the rest of Sweden'.[67] The campaign *Norrbotten Presents* created a register of emigrant Norrbothnians. The then County Governor has been reported to be positive about the register, primarily as it would consist of 'potential home-comers'.[68] The initiators of the Norrbotten Days obviously hoped to encourage Norrbothnians in exile to return to the region. Nevertheless, the project has been criticised as too costly, and the results of the event for being too meagre. A survey carried out by a regional daily paper indicates that only one single municipality had reported a 'home- comer' during the last year.[69] Partly because of the criticism, the project was ended in 2003.[70]

MECKLENBURG–WESTERN POMERANIA

In Mecklenburg–Western Pomerania (MV) we find several references to what is understood as a lack of human resources. Several voices do indeed argue for the sake of the elderly, the unemployed or other potentially disadvantaged groups; but, just as in Norrbotten, entrepreneurs and committed, charismatic and creative individuals are accorded extra attention in the regional debate.[71] A change in the regional mentality is in many contexts thought to be unavoidable if the region is to keep up with the global competition. The CDU maintains in a document from 1996 that the widespread *Subventionsmentalität*[72] constitutes a problem for the

66 Norrbotten Presenterar (2003).
67 Framtidslandet (2004)
68 NSD (2003b).
69 NSD (2003a), 'Fester för 7 miljoner, men utflyttarna vänder inte hem', 2003-06-12.
70 NSD (2003d), 'Slutfestat på Gärdet - kritiserat PR-projekt läggs ned', 2003-12-30.
71 CDU Landtagsfraktion MV (1996:18); CDU Landtagsfraktion MV (2002:5,20); PDS Landtagsfraktion MV (1999), *Was macht Ihr da im Schloss*, Schwerin; StK MV (2002c).
72 'Subsidy mentality.'

Land. When the strengths and the weaknesses of the *Land* are analysed in another CDU document a few years later, a 'palpable *Subventionsmentalität*', a 'missing entrepreneurial culture' and 'weak soft location factors' are brought out as weaknesses of the region.[73] The CDU views the alleged public tendency to rely on subsidies partly as a consequence of the GDR past. According to the CDU, people in MV must 'learn to live with freedom and democracy'. But this, it is argued, won't be easy, as 'many in the former communist dictatorships' prove to 'long for the old system, where the state cared for the citizens twenty-four-hours per day'.[74]

Other parties within MV rarely talk about *Subventionsmentalität* as a distinct feature of the people in the region. The PDS Minister of Regional Development, however, referred to research that claimed that the *Subventionsmentalität* and the aptitude for so-called *Massnahmenkarrieren*[75] are profound problems for the region.[76] A few years later, the very same minister referred to another report, said to indicate that 'awareness of professional independence as an alternative to wage earning' is particularly poorly developed in the north-eastern parts of Germany.[77] In a later comparison with other *Länder* in eastern Germany, the minister brings out the need for initiatives capable of increasing 'inventiveness' and 'entrepreneurial spirit' among the citizens in MV.[78] When a campaign to enhance the entrepreneurial spirit in the *Land* was launched in the early 2000s, it was stressed that the *Land* lags behind the rest of Germany when it comes to the establishment of companies.[79]

Brain Drain

Like Norrbotten, MV has suffered from negative migration flows during the 1990s and early 21st century. The migration from Mecklenburg–Western Pomerania is, however, more often perceived as a kind of 'brain drain'. That young

---

73 CDU Landtagsfraktion MV (1996:20); CDU Landtagsfraktion MV (2002:8).

74 CDU Landtagsfraktion MV (1996:8).

75 To do a *Massnahmenkarriere* is to get on in the world primarily by (ab)using the various subsidies that are provided by the social insurance system.

76 MfABL MV (2001a), 'Denkwerkstatt – Plädoyer für Mittelkonzentration', *Pressemitteilung Nr 229/01*, Schwerin.

77 MfABL MV (2003d), 'Unternehmergeist im Land vielfältig gefördert', *Pressemitteilung Nr 69/03*, Schwerin.

78 MfABL MV (2004a), 'Holter für eine Zukunft Ost in den Farben des Ostens', *Pressemitteilung Nr 24/04*, Schwerin.

79 MfABL MV (2003d).

educated people tend to leave the *Land* for employment elsewhere induced several parties to call for a policy to encourage these individuals to remain in the region. The CDU opposition communicates a firm belief that a certain kind of human resource would benefit the development of the *Land*. In a discussion on educational policy, it concludes that 'every society needs *Leistungseliten*'[80] and that the party aims to support these elites early on in life.[81] In another context, it is declared that the *Land* must prepare for the 'increasingly aggressive struggle for the *Beste Köpfe*'.[82] 'Without creative and well-educated minds', it is contended, MV will not be able to retain its prosperity, let alone enhance it. Therefore, it is proclaimed, '*wir müssen um die besten Köpfe kämpfen*'.[83]

While the CDU seems to be particularly persistent in bringing out 'the best' or 'the clever minds' as an important resource for the region, we find these arguments elsewhere too. In a reference to those working in biomedicine, biotechnology and agribusiness, the SPD President of the *Land* states that the 'true wealth of the *Land* is constituted by its *Köpfe*'.[84] Various representatives of the SPD stress that educated specialists make up a soft location factor, and emphasize that the importance of 'indigenous potential' and social capital for economic development has been confirmed by social science research.[85] The Minister of Regional Development refers to *Denkwerkstatt 2020* (see Chapter 6) as a gathering of '*kluge Köpfe*', capable of bringing new ideas to the region.[86] The PDS stresses in its election programme from 1997 that the *Land* must develop into an 'innovative and ... attractive "science location"', since this is an 'indispensable condition' for the future development of the *Land*. The Green Party stresses also that particularly young, well-educated individuals tend to leave MV. With them, it is said, 'we lose substantial potential for the development of the *Land*'.[87]

Along with calls for educated elites, knowledge and skills, we find calls for a new and more motivated attitude among the population. In the 1998 parliamentary address, all those who are 'motivated' and 'bear a willingness to work for

---

80 That is, people who belong to the upper stratum of society, not by birth or financial position, but by their achievements in society.

81 CDU MV (2002:25).

82 'The best minds'.

83 'We need to fight for the best minds'. CDU Landtagsfraktion MV (2002:20, emphasis in the original).

84 Ringstorff, Harald (2002), 'Vorwort', in StK MV (ed.) *Mecklenburg-Vorpommern – ein Land zum Leben*, Schwerin.

85 LandtagsNachrichten MV (2002:16); Müller, Wolfgang (2003), 'Stadstumbau Ost', *Horisonte* Nr 2/2003, p.39. See also SPD MV (2001:1).

86 www.denkwerkstatt-mv.de/pages/start.htm., 2004-06-13

87 Grüne MV(2002b).

the best interests of the *Land*' are invited to take part in the work developing the *Land*.[88] In a document from 2001, the SPD declares that 'our *Land* needs individuals who have the right attitude, and who are in a position to take part in the development of … the *Land*'.[89] The CDU for its part states that MV needs a '*generationsübergreifende Motivationsoffensive*'[90] and a new '*Aufbruchstimmung*'[91]. These features would be capable of 'giving us the strength and the self-confidence needed for coping with future prospects'.[92] A 'willingness to be *Selbständig*'[93] must evolve into a culture and a social movement in the *Land*.[94] This message is pretty much repeated in a document from the FDP, in which it is argued that Germany in general and Mecklenburg–Western Pomerania in particular need to advance 'a culture of independence and *Eigenantwortung*[95]'.[96]

The Entrepreneur

The private entrepreneur is in many contexts referred to as an important human resource. The PDS states in its election manifesto from 1997 that the party aims to protect and promote small and medium-size firms through a variety of policy measures.[97] A PDS representative in the *Landtag* confirms that 'support for and a strengthening of small and medium-size business' is part of party policy.[98] The PDS election manifesto from 2002 highlights support for 'innovative enterprises' once again. According to its manifesto, the PDS intends to offer 'extensive support to those who wish to establish a company, to small and medium-size firms and to artisans'. Western Pomerania – the poorer part of the *Land* – is said to be in particular need of a target-oriented campaign for attracting new and 'innovative companies' to the region.[99]

---

88 StK MV (1998:6,27).
89 SPD MV (2001:4).
90 Literally, an 'inter-generational motivational offensive'.
91 Literally, a 'feeling of upheaval' or 'a readiness to embrace change'.
92 CDU Landtagsfraktion MV (2002:5).
93 'Independent'.
94 CDU Landtagsfraktion MV (2002:9f)
95 'Self-reliance.'
96 Kreher, Hans (2003), *Liberaler Aufbruch MV*, Schwerin, p. 9.
97 PDS MV (1997b:3).
98 Borchart, Barbara (1999), in PDS Fraktion Landtag MV (ed.) *Was macht Ihr da im Schloss*, Schwerin.
99 PDS MV (2002:3f.).

A common assumption is that small enterprises need a certain social climate in order to become established and thrive.[100] In recent years, the coalition government has initiated a number of campaigns to create such a climate and to encourage people to set up their own companies.[101] The purpose of many of these initiatives has been to make private enterprise more popular, to 'support an entrepreneurial spirit, and to create a climate in which self-employment, innovation and commitment are brought to more general notice than before'.[102] In a press release from the PDS-led Ministry for Regional Development, it is proclaimed that Mecklenburg–Western Pomerania 'needs to become an enterprising *Land*'. 'We need more entrepreneurs and more enterprise in our *Land* . . . we need people capable of getting things done, and who dare to go for professional independence'.[103] The SPD President of the *Land* maintains in one of his announcements that the state cannot solve all problems, but that the 'vigour and the enterprising spirit of the citizens' will be essential for the future economic development of the *Land*.[104] The Minister for Regional Development contends in much the same manner that entrepreneurial spirit and setting up new businesses '*sind dem Land viele Millionen wert*',[105] and that the government aims to support these features with a number of new ways.[106]

The work of encouraging entrepreneurial spirit and creating a certain social climate involves campaigns and initiatives addressed to both students and schoolchildren.[107] A Social Democratic representative in the *Landtag* stresses that innovative individuals in general and young entrepreneurs in particular should be encouraged and promoted by the policy-makers.[108] Self-employment, according

---

100 PDS MV (2002:3); MfABL MV (2003d); Region Rostock (2004a) 'Rostock – Region der Chancen', Rostock.

101 One example of such a campaign is 'Einfach Anfangen' ('Just Start!') financed jointly through MfABL and the European Social Fund. The campaign aims primarily at 'supporting the entrepreneurial spirit in the *Land*' MfABL MV (2003c), 'Prominente unterstützen Initiative für Unternehmergeist', *Pressemitteilung Nr 56/03*, Schwerin.

102 MfABL MV (2000c), 'Kräftiger Impuls für Existensgründer', *Pressemitteilung Nr 222/00*, Schwerin; MfABL MV (2000d), 'Unternehmer braucht das Land – nicht Unterlasser', *Pressemitteilung Nr 225/00*, Schwerin.

103 MfABL MV (2000d).

104 StK MV (2003b), 'Mit Zuversicht und Selbstbewusstsein in die Zukunft sehen', *Pressemitteilung Nr 222/03*, Schwerin.

105 'is worth many millions to the *Land*.'

106 MfABL MV (2003d).

107 MfABL MV (2000d).

108 LandtagsNachrichten MV (2003c), *Quoting Schildt, Ute*, 6/2003, Schwerin, p. 5.

to the Minister for Regional Development, needs to be a matter of course also for the younger generation.[109] A series of press releases from the ministry stress that private enterprising implies increased opportunities for youngsters to stay in the *Land* after having obtained their degrees. Given that, it is maintained, a 'new and more tolerant attitude towards private enterprises' must be integrated into schoolteaching.[110]

The Role Model

Both the opposition and the government seem convinced that role models and 'good examples' are central to the work task of encouraging self-employment in the *Land*. The CDU asserts that an 'improvement of the image of the private entrepreneur' needs to take place, and that this is something that demands personalized initiatives and role models.[111] The Ministry for Regional Development declares that it is working for 'more courage to seek self-employment' partly by promoting a more positive image of the private businessman'.[112] The idea that 'entrepreneurs should recruit entrepreneurs'[113] is fundamental to *Einfach Anfangen!*,[114] an initiative taken by the *Land* government. Prominent businessmen from the region have been asked to relate their 'individual success-stories' in a large-scale advertising campaign designed to infuse 'more courage to seek professional independence or self-employment' in the region.[115] The rationale of the decision to promulgate these success stories to the inhabitants is that people won't be willing to establish their own companies unless they are convinced that 'it works'. Accordingly, the campaign highlights 'those who set a good example' to the rest of the *Land*. By using 'real' and 'authentic' examples from within the region, the campaign will allegedly contribute to a new climate

109 MfABL MV (2000d).

110 MfABL MV (2000b), 'Land braucht gutes Klima für junge Unternehmer, *Pressemitteilung Nr 105/00*, Schwerin; MfABL MV (2000d); MfABL MV (2003e), 'Unternehmen sein ist toll – Unternehmen bleiben toller', *Pressemitteilung Nr 145/03*, Schwerin.

111 CDU Landtagsfraktion MV (2002:29).

112 MfABL MV (2002a), 'Kampagne für Gründer', *Pressemitteilung Nr 214/02*, Schwerin.

113 *Gründer werben für Gründer* in the original.

114 'Just do it!'.

115 MfABL MV (2003c); MfABL MV (2002a); http://kampage.einfachanfangen.de/kampagne/kamp.html, 2004-11-08.

in which 'entrepreneurial spirit' and 'courage to ... choose self-employment' becomes natural for the inhabitants in MV.[116]

The MV Diaspora

Just like Norrbotten, Mecklenburg–Western Pomerania hosts a project to attract 'home-comers' to the region. The project – *MV4You* – was initiated by the Ministry for Regional Development in 2001, and is described as having those who have left the *Land* for better opportunities in other parts of Germany as its main target group. Occasionally, it is stressed that *MV4You* should be an interface not only between employers in the *Land* and potential home-comers but also between employers and potential immigrants from other parts of Germany. According to the Minister for Regional Development, the work with *MV4You* proves that there is a 'clearly recognizable interest in returning to or immigrating to Mecklenburg–Western Pomerania'.[117]

Still, it is the potential home-comers – those who have left Mecklenburg–Western Pomerania due to unemployment – who stand out as the main target group. Government representatives have in a number of contexts stressed that people who have left MV should be contacted and be given the chance to return to their *Land*.[118] The PDS group in the *Landtag* argues that *MV4You* should 'keep in contact with emigrants, in order to win them back to the *Land'*.[119] In general, the government communicates the idea that people may indeed leave MV for further qualification elsewhere, but that this group of people must be encouraged to return to their former homeland. In actual fact, the motto of the *MV4You* initiative has been that it is 'good to move and to gain experience' and 'good to return to the *Land* too'.[120] Accordingly, *MV4You* is understood, not as a general recruitment office, but as an agency that is to 'keep in contact with those who have left Mecklenburg–Western Pomerania because they have no

116 http://kampage.einfachanfangen.de/hintergrund/hintergr_idee.html, 2004-11-08.

117 MfABL MV (2003a), 'Wandern und Wiederkommen - eine Chance fur das Land, *Rede des Arbeitsminister Holter*, 2003-03-26, Schwerin.

118 MfABL MV (2001c), *Arbeitsmarkt- und Strukturentwicklungsprogramm*, 2001-12-04, Schwerin; see also Landtag MV (2003b), Holter, Helmut in *Plenarprotokolle*, 4/10, 12. März, Schwerin; Landtag MV (2002b), Holter, Helmut in *Plenarprotokolle*, 3/81, 29. Mai , Schwerin.

119 PDS Landtagsfraktion MV (2005), *Presseerklärung No 707*, 2005-04-19, Schwerin.

120 One of the slogans used by *MV4You* has been 'wandern und wiederkommen'.

professional opportunities in the *Land* at the moment'. Hence, 'those who leave should not be forgotten'.[121]

Clearly, well-educated specialists constitute an important target group for *MV4You*. The agency is said to keep in touch with what it calls 'highly qualified people willing to immigrate or return to Mecklenburg–Western Pomerania'.[122] The Minister for Regional Development maintains in a speech that 'many young specialists, after some years working in other parts of Germany, feel the desire to return to their homeland'. This, it is said, is 'a great chance for MV'.[123] The importance of keeping and bringing 'clever minds' to the region was stressed also in a later campaign, *Kluge Köpfe für MV*', in which people who are willing to work against the 'emigration of intelligence' are asked to contribute with their ideas and their know-how.[124] The agency has indeed been criticized by the CDU in the *Landtag*, as well as by the FDP in the *Land*, for being inefficient and costly.[125] Individual PDS representatives have also questioned the cost associated with *MV4You*.[126] Yet the very idea behind the strategy – namely, that 'home-comers' constitute an important target group – enjoys wide support.

A Civic Regionalism?

In the definition of regionalism adopted here, the region's population accords a central position.[127] In Chapter 7 we encountered a rather romantic picture of this population. It was widely assumed that the regional population shares some common characteristics, a common cultural heritage, and a common identity. In view of that, the regionalism revealed here might be suggestive of a romantic or cultural nationalism, characterized by what Calhoun has called a propensity to stress the 'ethnic or cultural similarity of the members of a political commu-

---

121 www.am.mv-regierung.de, 2005-01-21.

122 MV4You (2005a), 'Projektbeschreibung' at *www.mv4you.de/project_description.php*, 2005-05-19.

123 MfABL MV (2003a); See also MfABL MV (2001c).

124 MV4You (2005b), 'Kluge Köpfe für MV' *at* *www.klugekoepfe4mv.de/uploads/flyer.pdf*, 2005-07-28.

125 Landtag MV (2003d), Rehberg, Eckhardt in *Plenarprotokolle*, 4/13, 10. April, Schwerin; Landtag MV (2004a) Riemann, Wolfgang in *Plenarprotokolle*, 4/31, 18. Februar, Schwerin; Klink, Harry (2005), *FDP gegen Hochschulsterben - Hochschulen rechnen sich!* PM vom 28.04.2005.

126 Landtag MV (2003a), Gramkow, Angelika in *Plenarprotokolle*, 4/13., 10. April, Schwerin.

127 Williams (1997:114).

nity'.[128] The exposition in the present chapter, however, moderates any such expectation. Here, the (potential) entrepreneur and the well-educated are brought out as highly important sections of the population. Accordingly, there seems to be a strand of 'civicness' in the regional policy discourse too, even if policy thinking in Mecklenburg–Western Pomerania and (particularly) Norrbotten is far from 'neutral' towards ethnic or cultural belonging.[129] The regional population is occasionally lumped together and ascribed with some supposedly inherited mental assets or deficits. Besides, we do find a willingness to entice primarily 'home-comers' back to the region. Just like *MV4You*, Norrbotten Days addresses many of its messages to those who have emigrated from the regions. Yet 'new-comers', capable of contributing to the development of the region, are also welcomed. The regional citizen is, therefore, not defined in terms of origin only, but also in terms of what he or she would bring to the region in terms of economic development.

Employability and Human Resources

In academic debate, shared norms, trust and reciprocity are recurrently assembled under the concept of *social capital* – a concept mostly associated with Robert D. Putnam and his *Making Democracy Work*. Politicians, as well as scholars from a wide range of disciplines, have referred to Putnam's study when arguing that regional development depends on civic virtues and customs. Besides Putnam, a great many other theorists agree that 'conventions of dialogue, trust and reciprocity, and, in some local cases, a culture of social and civic solidarity'

---

128 Calhoun, Craig (1994), 'Nationalism and Civil Society: Democracy, Diversity and Self-determination', in Calhoun, C. (ed.), *Social Theory and the Politics of Identity*, Oxford: Blackwell, p.305.

129 According to Kymlicka, the ideal typical civic nation has been perceived as 'liberal' in the sense that it takes a 'neutral' line towards the ethnic or cultural identities of its citizens. But this interpretation has met severe and growing criticism. Kymlicka, too, makes a strong case against the division between romantic and civic nationalism in Kymlicka, Will (2000), 'Estonia's Integration Policies in a Comparative Perspective', in *Estonia's Integration Landscape: From Apathy to Harmony*. Tallinn: Jaan Tonissoni Instituut, pp. 29-57. See also Eisenstadt, S. N (2002), 'The Construction of Collective Identities and the continual reconstruction of Primordiality', in Malesevic, S and Haugaard, M. (eds.), *Making Sense of Collectivity*, London, Sterling, Virginia: Pluto Press.

are variables that can explain economic success.[130] All in all, interest in the social foundations of economic development has grown considerably during recent decades.[131]

The focus of works inspired by Putnam is often on informal societal structures. There is a palpable tendency to praise 'social capital', 'trust' or other collective resources. This approach, however, has also had repercussions on the understanding of the *individual*. There is, as manifested in the empirical sections above, a propensity to pay a certain tribute to individual skills and resources such as entrepreneurialism, creativity, will, and education. This propensity can be linked to a wider discourse on 'employability', a concept that has become a buzzword in European labour market policy discussions.[132] Hence, 'employability' is thought to include 'the collection of workers' characteristics ... that are increasingly seen as determinants of employment chances'.[133] Examples of characteristics thought to increase individuals employability are 'positive attitudes towards work', 'adaptability', 'service-mindedness' and 'flexibility', but also 'skills' and 'an entrepreneurial attitude'.[134]

Almost all of the qualities cited above are mentioned in discussions about regional growth and development. Peoples' spirit of enterprise, ways of thinking, attitudes, forms of social intercourse, norms and values, have all been cited to explain regional economic success and failure.[135] The alleged importance of

---

130 Amin, Ash (2000), 'Industrial Districts', in Sheppard, E. and Barnes, T. (eds.), *A Companion to Economic Geography*, Oxford: Blackwell.

131 For an overview, see Boschma, R.A. (2005), 'Social Capital and Regional Development: An Empirical Analysis of the Third Italy', in Boschma R.A. and Kloosterman, R. (eds.), *Learning from Clusters: A Critical Assessment*, Dordrecht: Springer Verlag. See also Colletis-Wahl and Pecqueur (2001:450).

132 Garsten, Christina and Jacobsson, Kerstin, eds. (2004), *Learning to be Employable: New Agendas on Work in a Globalizing World*, Basingstoke: Palgrave Macmillan. See also McQuaid, Ronald W. and Lindsay, Colin (2005), 'The Concept of Employability', *Urban Studies*, 42(2), pp.197-219; Green, Anne E. and Turok, Ivan (2000), 'Employability, Adaptability and Flexibility: Changing Labour Market Prospects', *Regional Studies*, 34(7), pp.599-600; Peck, Jamie and Theodore, Nikolas (2000), 'Beyond "'Employability'"', *Cambridge Journal of Economics*, 24(6), pp.729-749.

133 Peck and Theodore (2000:731).

134 Garsten and Jacobsson (2004), Green and Turok (2000:599-600), Peck and Theodore (2000).

135 For some additional Swedish examples, see Västra Götalandsregionen (2004), *Regionalt Tillväxtprogram för Västra Götaland 2004-2007*, Göteborg; Lst Värm-

employability can be seen in obligatory vocational training courses or, as in our case, in campaigns for entrepreneurship and self-employment. Irrespectively of how it is manifested, however, discourses of employability tend to proceed from the idea that unemployment is the result of deficits on the supply side of the labour market.[136] The great importance attached to employability has been said to illustrate a shift of focus from demand-side labour market factors to a 'focus on individuals and their qualities'.[137] As such, it has been understood as the opposite to Keynesianism, which aimed rather at supporting the demand side of the economy.[138] Therefore, employability has in more than one context been linked to liberal or so-called Third Way politics.[139]

This shift of focus from the demand side of the labour market to individual characteristics involves a more palpable stratification of the workforce. Recent narratives of growth tend to write out 'those with the skills and experience which are considered inappropriate for the worlds of high-tech business and private services'. The 'physical strength and industrial skills of the mass collective worker' are, as argued by Allen *et al.*, increasingly seen as redundant.[140] Moreover, the stress, particularly in MV, on *'kluge Köpfe'*, highly skilled expertise and know-how has an embedded class dimension. Thus, children of university-educated parents are generally more likely to enter higher education than working-class children. In Germany, this is even more significant. International comparisons have shown that in Germany a relatively small share of the younger generation is enrolled in university education. More important, however, is the fact that children from higher socio-economic backgrounds make up a much greater proportion of the student body in Germany than in other OECD countries.[141]

The policy discourse do not draw attention to the class dimension embedded in the celebration of well-educated specialists or entrepreneurs. It is more com-

---

lands län and Region Värmland (2003); Lst Västernorrland (2001), *Vision 2005 med perspektiv mot år 2010,* Västernorrland; Östsam (2000).

136 McQuiad and Lindsay (2005:204), Peck and Theodore (2000:729).

137 Garsten and Jacobsson (2004:2).

138 Tomaney and Ward (2000:473).

139 Doyle, Mike (2003), 'Discourses of Employability and Empowerment: Foundation Degrees and "Third Way" Discursive Repertoires', *Discourse: Studies in the Cultural Politics of Education*, 24(3), pp.275-288; McQuiad and Lindsay (2005).

140 Allen *et. al* (1998:100).

141 Usher, A. and Cervenan, A. (2005), *Global Higher Education Rankings 2005.* Toronto: Educational Policy Institute; Usher, A. (2004), *A New Measuring Stick. Is Access to Higher Education in Canada More Equitable?* Toronto: Educational Policy Institute., p.8.

mon for expert knowledge and entrepreneurship to be presented as accessible for everyone. The underlying logic is that anyone can be a successful entrepreneur or expert as long as the willpower is there and as long as there are role models in the region capable of proving that 'it works'.

The Politics of Virtue

It is, to say the least, a rather one-sidedly optimistic account of entrepreneurship that stands out in regional policy discourse. This is, however, not exceptional, either for the regions studied here or for the regional level in general. The state level in both Sweden and Germany tends to communicate the idea that increased entrepreneurship and vigorous enterprise are the main path to regional economic development. Besides, research on entrepreneurship and self-employment, too, tends to be indulgent towards the unpleasant sides of entrepreneurship. More often, it is contended that 'increased entrepreneurship and lively free enterprise' is the 'only chance for survival in many regions'.[142]

One plausible explanation for the lack of research on the *problems* associated with small-scale enterprise and entrepreneurship is, as Rehn puts it, that most research on entrepreneurship proceeds from the normative assumption that entrepreneurship is 'a good thing'.[143] The possible underside of entrepreneurship (which according to Rehn could be stress-related illnesses, huge personal overwork, poor returns and the risk of bankruptcy) are paid little attention, either in politics or in research.[144] To be entrepreneurial has to all appearances turned into a *virtue*.

In fact, the whole discussion about how the regional citizen should behave seems to proceed from a rather shallow *ethics of virtue* perspective. Thus, the notion that the regional citizen should change her attitudes and manners in a number of ways could be linked to a discussion about 'human ideals' or 'good human beings'. A virtue can be described as an 'acquired trait of character' or as a 'manifestation of an ideal, existing in a certain social context'.[145] Some virtues are of a general kind, whereas others are valid only in a certain sphere. There are

---

142 Nilsson and Brulin (1997:5).

143 Rehn, Alf (2004), 'Pessimistiskt Entreprenörsskap', *The Pink Machine Papers*, 1, Stockholm.

144 Rehn (2004).

145 Aronsson, P., Aspers, P., Beckman, L., Erixon, F., Micheletti, M., Möller, G., Strömberg, K., Uddhammar, E. (2001), *Dygder som drivkraft och föredöme*, Stockholm: City University Press, p.24.

virtues that are seen as desirable in private life, whereas others are sought rather in public life. In public life, it is the *civic virtues* that are of interest. In the republican tradition, it has been argued, a central civic virtue is participation in political life. Voting in general elections, taking part in public debate, paying taxes or caring for the environment have also been mentioned as central civic virtues. In working life, qualities such as loyalty, honesty punctuality and politeness have been brought out as virtues worth aiming at, even if ideas about what a virtue is have varied between lines of business and places of work.[146]

In today's debate about regions and regional development, however, other virtues than those mentioned above come to the fore. Besides, the relativism embedded in an ethics of virtue perspective implies that old or existing virtues can be renegotiated and endowed with new meaning. Hence, virtues that used to have democratic or emancipatory connotations can now be reinterpreted and given a more economistic undertone. In this context, Garsten and Jacobsson point out that 'human capabilities' and 'knowledge' are 'considered more important than ever to economic and social welfare'. Given that, features such as 'learning' and 'education' are no longer given the emancipatory stress they had before. Instead, it is argued, these features now refer to the 'socio-economic need for mobilizing and adapting human resources in the struggle for economic growth, productivity and competition'.[147] Thus, apart from virtues such as steadiness and reliability, there is an increasing focus on qualities that are virtues in the light of their alleged ability to support economic development. It is, quite generally, virtues that are thought of as favourable for economic growth that are praised and given flattering descriptions.[148] Besides, alleged deficiencies such as 'moaning' or 'passivity' are seen as such primarily because of their alleged capacity to counteract regional competitiveness. Such deficiencies are then condemned and met with attempts to 'enlighten' and sophisticate the population. This enlightenment is thought in both regions to be integrated into schoolteaching, and to be communicated to the population through a variety of campaigns.

Apparently, the economic bias of contemporary regionalism does affect the idea of what the central civic virtues are. This is central, since *civic nationalism* has from time to time been conceptually associated with struggles for democracy and human rights. Civic nationalism has in these cases been understood as a civilized opposite to supposedly backward or 'fundamentalist' ethnic nationalism. This understanding of civic nationalism has, however, been severely criticized, as the experience of centuries of nationalism has shown that most forms of

146 Aronsson *et. al.* (2001:35-38).

147 Garsten and Jacobsson (2004:2).

148 Aronsson *et al.* (2001:15,73).

nationalist thought involve large-scale exclusion and discrimination.[149] The position taken here is that most nationalist – and regionalist – doctrines most probably have strains of both civicness and more ethnically or culturally loaded arguments and practices.[150] Accordingly, we ought to ask how these civic or cultural strains are manifested in each case. To judge from the empirical exposition above, the civicness of contemporary regionalism is not primarily manifested in a desire to broaden or deepen the human rights of citizens. Economic growth and development, rather than formal democratic rights, constitute the universal, civic or non-ethnic aims of the kind of regionalism displayed here.

Laggards and Forerunners

In studies on nationalism and nationalist policy, the discursive and social construction of 'us' and 'them' has been an object of a general interest.[151] 'Othering' has thus been understood as a way of 'defining and securing one's own positive identity through the stigmatization of an "other"'.[152] The idea of the self versus the other has, as Paasi suggests, 'become pivotal to modern social theory'.[153] What we find in our regions, however, is a very elaborate 'we' with no evident or explicit corresponding 'other'. The alleged disposition of the regional *self* is paid a lot of attention, whereas accounts of the outsiders or the non-belonging 'others' are pretty hard to find. Indeed, one could contend that the historical references, the talk about roots and belonging *per se* exclude those who cannot claim to have their roots and their origin in the region. Yet this 'othering' is very much implicit in the way that it pays no attention whatsoever to those excluded from the regional community. The dichotomies between 'us' and 'them', 'friends' and 'enemies' simply do not stand out as the most central ones in regional policy discourse.

---

149 Calhoun (1994:311).

150 Kymlicka, Will (1995), 'Misunderstanding Regionalism', *Dissent*, Winter, p.130-135.

151 Hylland-Eriksen (1999:19); Saïd, Edward, W. (1993), *Orientalism*, Stockholm: Ordfront; Hall, Stuart (1997), *Representation: Cultural Representations and Signifying Practices*, London: SAGE; Harvey, David (1993), 'Class Relations, Social Justice and the Politics of Difference', in Keith, M.and Pile, S. (eds.), *Place and the Politics of Identity*, New York: Routledge, p.63; Billig (1995); Keating (1998:90).

152 Ullrich, Melanie (2003), 'Definitions of Othering' at *www.cwrl.utexas.edu*, 2004-11-26; Crang, Mike (1998), *Cultural Geography*, London: Routledge, p.60.

153 Paasi (1996:13); Löfgren (1989:11).

The narratives referred to in this chapter are in fact less structured around the spatial boundaries of the region than around a time scale. The 'past' thus represents something rather positive, but increasingly outmoded. The 'present' is recurrently conceptualized as a turning point, as a juncture at which the regional mindsets, habits, attitudes, and manners need to be changed. The 'future', to conclude, is in both regions perceived as a time full of opportunities, but also as a time that demands new virtues and new patterns of behaviour. This metaphorical timescale could be used also to describe the ways in which regional citizens are represented. The regional citizen is thought to be in a process of change, and accordingly there are no absolute virtues or no absolute 'rules of conduct, traditions, or social routines that define and demarcate the boundary of the collective'.[154] Accordingly, the most patent trait in the regional policy discourse is not to welcome people according to their inclination to conform to the *old,* inherited conventions. Instead, the main stress is on those capable of adopting *new* habits, *new* routines, or a *new* way of thinking. Figuratively speaking, the regional policy discourse is not structured only around insiders and outsiders. Instead, it is construed also around categories such as *forerunners* and *laggards*. The forerunners in the regional policy discourse are largely made up of entrepreneurs and role models. The entrepreneurs and those with the right will, attitude and commitment, are the ones that are thought to bring economic growth and development to the region. Another patent group of forerunners, particularly in MV, is constituted by well-educated specialists. In both regions, however, we meet with a desire to make the regional population more interested in knowledge and education and to entice high-skilled competence to the region. The laggards in this discourse are obviously those who are uneducated or unemployed, who hesitate to start their own companies, or who simply lack the right attitude.

The individualistic bias stands out clearly here. The regionalism found in Norrbotten and Mecklenburg–Western Pomerania is not necessarily a policy for a particular cultural or territorial community. Instead, it is a policy that could be understood as modelled for a particular type of individual, namely, the motivated, enterprising and successful one. Given this, the most plausible target group for the kind of regionalism analysed here would consist primarily of those who can identify themselves *both* with the essentialized image of the citizen, referred to in Chapter 7, *and* with the image of the committed entrepreneur or the experienced, well-educated specialist. Those who lack the right roots, who happen to live on social allowances, who hesitate to start their own enterprise or who lack the appropriate education may be less attracted by its content.

154 Eisenstadt (2002:37).

Conclusion

The aim of this study has been to investigate what kind of regionalism is emerging in less favoured, politically defined European regions today. More precisely, the ambition has been to understand the *value basis* of such a regionalism. Throughout the work, it has been argued that regionalism can be understood as a political ideology, and thus observed and analysed through an examination of political discourses. It has also been argued that the study of regionalism can, preferably, be structured around some dimensions that can be found in many other political ideologies. Therefore, the study has focused upon 1) the world-view, 2) the core values, 3) the organizational ideals, 4) the means for mobilization, and 5) the target group of regionalism. This mode of procedure provided the study with a more 'theory-driven' approach, and was elaborated in order to facilitate a more systematic and less arbitrary 'discovery procedure'.[1]

In the first part of this concluding chapter, I summarize what I see as the central empirical observations. In the intermediate section, I put forward five concluding arguments, comprising what I consider to be the most important theoretical contributions of the study. Lastly, I discuss some questions relevant to future research on regionalism and region-building in a European context.

Dimensions of Regionalism: Empirical Findings

The empirical study evinced that the *world-view* of regionalism, as manifested in the two regions of Norrbotten and Mecklenburg–Western Pomerania (MV), is predominantly optimistic. Political actors in both regions tend to represent the current world order as essentially positive for the regional level. The world is indeed referred to as both 'globalized' and 'Europeanized', but it is also said to be less and less dominated by central states. In both regions, the state is described as an institution that is losing more and more of its power. In this context, it is recurrently maintained that regions are becoming increasingly important – for the economy and politics as well as for cultural- and identity-related matters. Globalization, Europeanization and declining state power are all interpreted as processes that shift power to the regional level and that further the requirements of regional competitiveness. In Norrbotten, however, such

1 van Dijk (2004:14).

arguments about regions gaining in political power occur more frequently than in MV.

The empirical review also demonstrated that political representatives in both regions regard the economy as centrally important. Clearly, global competition and the alleged dismantling of the central state have turned economic growth and competitiveness into a *core value* at the regional level. It is not 'congruence between political and cultural spaces'[2] or some sort of 'self-government'[3] that stands out as the regionalist keystones. Recognition, previously conceived as 'the paramount objective of nationalist movements in nations without states',[4] does not really dominate the policy agenda either. The striving for recognition is certainly not without significance, but it is not recognition primarily for culturally distinctive features that is sought. Neither is it recognition from the *state* or from the international political community that regional representatives are primarily aiming at.[5] Instead, it is recognition for being an attractive place for investment that is the focus. It is recognition for the region from its own population, from tourists, from potential settlers, from foreign investors and from the business community at large that regional representatives want. One of the most common means for achieving that recognition is image campaigns.

The regionalism identified is obviously an ideology driven by economic arguments. Besides economic growth, however, regional unity has a central status. Unity is thought to be beneficial for economic growth and competitiveness, and could therefore be regarded as a means to an end. Unity, however, also stands out as an end in it self. The ideal region is represented as 'united', and as a space in which all citizens 'pull together', adopt the same *Leitbild*, and put disagreements aside for the sake of the common, regional, good.

When the *organizational ideals* of regionalism were scrutinized, it was found that working in networks and partnerships is seen as the most appropriate political strategy. Global competition is thought to expose the region to new challenges; and, in order to cope with them, it is maintained, regional representatives must 'learn to cooperate'. Ideological dividing lines and organizational boundaries are, accordingly, represented as 'barriers' that need to be torn down. In both regions, we saw examples of partnerships and development coalitions being formed in order to create consensus in regional developmental work. Whereas several such partnerships were found at the

2 See Gellner (1997:11); Gellner, Ernest (1999), *Nationalism*, Nora: Nya Doxa, p. 24.
3 Kellas (1998:7),
4 Guibernau (1999:25),
5 See Guibernau (1999:25),

regional level in Norrbotten, we found them primarily at the subregional level in Mecklenburg–Western Pomerania.

Even if 'culture' or cultural recognition fail to qualify as a core value, issues related to culture still play a central role. Cultural traits presumed to be 'authentic' or 'genuine' are used as *means for mobilization*, but are also seen as important for the 'institutional milieu' in the region. Traditional, place-bound cultural expressions are in both regions thought to support the emergence of a solid regional identity. They are also represented as something that could be commodified and marketed in the so-called cultural economy. Besides, culture plays a role in regional image strategies, as a means of attracting public support and loyalty as well as investors and tourists to the region. In that respect, culture is certainly used as a means for mobilization, and in the quest for recognition. Still, it is not culture for culture's sake that is in focus, but its alleged advantages for the regional economy.

'Genuine' or 'traditional' cultural expressions are obviously thought to be advantageous for the economic development of the region. This notion, however, is forwarded only up to a point. Thus, the target group of the kind of regionalism identified here does not necessarily consist of people who have the 'right' origin. A firm attachment to the region is indeed welcome, and to 'have roots in the region' is certainly seen as an advantage. People who once left the region attracted by better prospects elsewhere are also seen as a potential resource, who must be brought back to their 'native land'. Romanticist, essentialist accounts of 'the regional population' are thus common, particularly in the Norrbottnic case. Yet the allegedly 'regional' traditions and habits are not treated as entirely beneficial. Some aspects of 'the regional way of life' are dismissed as an obstacle to the transformation of the region into a modern and competitive space. Instead, it is those with the right skills and the right attitude that are most commonly celebrated in regional policy discourse. In both regions, 'entrepreneurs', 'role models' and 'experts' were paid a great deal of attention.

Theoretical Contributions

The observations recounted above can be of use in further developing theories of regionalism and region building. Below, I offer five concluding arguments that either complement or question previous ways of thinking about both regionalism and region building. *First*, I argue that my study establishes that definitions of regionalism cannot focus solely on claims for autonomy or calls for recognition. The value basis of regionalism can apparently vary over time as well as over space and, therefore, we need definitions of regionalism capable of acknowledging that variability. *Second*, I point out that the regionalism identified has strains of both romanticism and civicness. Therewith, I claim to nuance the most dichotomizing accounts of romantic ethno-regionalism on the one hand and supposedly civic, economically driven forms of regionalism on the other. My *third* argument is that the regionalism found – in its romantic as well as in its civic aspects – is involved in a process of region-building. It accentuates the outer boundaries of each region, it forwards a certain image of regional history, and it is engaged with the formation of a rather exclusionist regional community. My *fourth* argument is that representatives of regions, states, the European Union, and the regional sciences are involved in a *thought collective*,[6] sharing a certain way of thinking about regions and regional development. This way of thinking, I suggest, is an important 'horizontal' trend that fuels a certain kind of regionalism. My *fifth* concluding argument is that some regions may be more likely than others to become 'regionalist'. In this context, I stress that ethnicity and culture can play a central role also in administrative regions – which are not known for hosting 'ethno-regionalist movements' or for being distinguished by certain allegedly pre-existing, cultural characteristics.

Accurate Definitions of Regionalism

The *first* thing we can learn from this study is that regionalism cannot be defined only by reference to 'claims for autonomy', 'feelings of regional belonging' or 'calls for cultural recognition'. Thus, the value basis of regionalism is not predetermined but changes according to its context. This study has clearly demonstrated that claims to, and calls for, autonomy or cultural recognition are not the most central ones. Importantly, however, this study has focused solely on what is understood as a political elite. Other sources and informants from within the same regions may well have provided another kind of regionalism. For that reason, we need definitions of regionalism that are capable of accommodating the complexity and the elusiveness of regionalism. The definition forwarded by Williams[7] has this capacity, it has been argued, since it allows us to identify different kinds of regionalism in different time and place contexts.

---

6 Fleck (1997).

7 According to Williams (1997:114), regionalism can be defined as 'the attempt to optimise the interests of a region's population through the manipulation of the political process'.

The fundamental assumption that regionalism can take different shapes in different settings is a weighty argument against attempts to 'measure' or 'survey' regionalism or regionalist mobilization by using statistical data analysis only. Since regionalism proves to be so complex in nature, we cannot presume that there are any universal 'indicators' of regionalism. Subsequently, attempts to predict the rise of regionalism in a large number of dispersed cases by plotting some predetermined, measurable indicators will prove to be essentially problematic. The complexity and variability of regionalism speak, rather, in favour of using qualitative methods to explore and analyse a small number of cases.

The Reciprocity of Romantic and Civic Traits in Regionalism

My *second* argument relates to the fact that the regionalism identified has both romantic and civic strains. On the basis of that observation, the study tends to argue against any notion that recent forms of regionalism should be different from earlier ones by virtue of their 'openness'[8] or lack of 'virulent ethnic exclusionism'[9].[10] The aforementioned economic bias, the ambition to work in partnerships, and the accent on entrepreneurs and highly skilled experts all indicate an affinity with the kind of regionalism that emerged in western European growth regions in the late 1980s. This 'new regionalism' has in turn been understood as being both civic and cosmopolitan, and as less reactionary, backward or romantic than previous forms of regionalism. Empirical examination, however, establishes that the regionalism identified is far from 'neutral' towards ethnicity or towards different cultural expressions.[11] We find a wealth of references to cultural history and folklore, regional languages, dialects and identities. Regional, place-bound identities and cultural peculiarities are also accorded far greater significance than non-regional identities and cultural expressions. The argument, as forwarded in Chapter 7, was that the celebration and commodification of allegedly authentic regional cultures involves a stratification of different cultural expressions.

The celebration of regional culture and identity can scarcely be understood as a reaction against national or 'European uniformity', 'Americanization' or an

8 Törnqvist (1993:42).
9 Schmitt-Egner (2002:188).
10 cf. Gren (1999:30ff; 2002b:19,89); Gidlund and Sörlin (1993:218).
11 See, for example, Kymlicka (2000).

alleged global, cultural assimilation.[12] The particular is defended, not for its own sake, but rather because of its alleged capacity to stimulate the regional economy.[13] The genuine is said to benefit regional development in a direct way through the cultural economy or the tourism sector, or indirectly by creating 'social capital', 'feelings of togetherness' or other 'soft development factors' in the region. Besides, notions about regional identities and 'mentalities' come to the fore not only in discussions about culture or cultural heritage. True, in the political discourse, it certainly appears as if 'identity' can be thought of and represented only in terms of cultural and territorial belonging.[14] Still, notions about identities, 'mentalities' or 'regional ways of being' come to the fore also in discussions about people's (dis)inclination to contribute to regional economic growth. Obviously, arguments about culture, cultural heritage, identities and mentalities are not separated from ideas about competitiveness and growth. Instead, civic and romantic notions are intrinsically intertwined with one another and thus capable to coexist in the same value system.

Civic and Romantic Elements in the Building of Regions.

Having demonstrated that civic and romantic notions can coexist in the same value system, my *third* argument is that romantic and the civic notions alike host palpable region-building ambitions. Thus, the regionalism identified accentuates the outer boundaries of the region. It promotes a particular, official, regional history and is concerned both with the formation of a regional community and with the construction of a particular kind of regional institution. The emphasis on allegedly historical, cultural boundaries has romantic or culturalizing overtones. The stress on common regional interests and joint action reflects, instead, a civic ideal.

(a) Boundary Formation

In Norrbotten as well as in Mecklenburg–Western Pomerania, regional representatives do talk about the many influences from beyond the political borders of the region. In Norrbotten, the influences from Finland and from the border-straddling Saami minority are very much referred to. In MV, the geographical closeness to Scandinavia and the historical contacts to the countries

---

12 See Guibernau (1999:23); Löfgren (1989).

13 Keating (1997:30).

14 For a discussion on this, see Hansen (2000:179).

around the Baltic Sea are recurrent topics in the regional policy discourse. This implies that the political borders are both crossed and defused in political discussions about history and culture.

The regionalism identified is, nonetheless, conducive to the construction of boundaries. The social construction of boundaries can thus take many forms. Here, it is particularly the outer boundaries related to cultures, identities and emotions that are stressed, whereas inner boundaries are either neglected or referred to as a problem. We find, particularly in Norrbotten, an 'exaggerated account of homogeneity within the borders'[15] and an augmentation of dissimilarities with the outside. Thus, Norrbotten is referred to as a unique mixture of three allegedly distinct and place-bound cultural traditions – the Swedish, the Saami and the Finnish. In MV, the boundary between Mecklenburg and Western Pomerania is intermittently acknowledged, and the 'distinctive features' of the two parts of the *Land* are also considered. Besides, there are four administrative subregions in the *Land*, in which efforts are made to promote their distictiveness. Still, the desire to forward a positive and coherent image of the entire *Land*, and to mobilize wider sections of the regional population around it, also implies a highlighting of the outer boundaries.

The study has confirmed that in politics as well as in academic work there is a disinclination to problemize or to contest the *outer* boundaries of the region. A yet more central argument, however, has been that there is very pronounced tendency – in politics and in academic work alike – to engage with and to meticulously plot the *content* of the region. The focus on endogenous potential has motivated countless attempts to specify the status of the human resources, the social capital, and other 'soft development factors' in the region. This engagement with the content of the region, I have argued, also constructs boundaries. Thus, to bring a multitude of social phenomena, cultural traditions and traits together, and to label them as being 'typical' for a particular territory, is certainly an act that stresses the outer boundaries of that territory.

The ambition to create common regional stances was patent in both Norrbotten and MV. This ambition, too, enforces the outer boundaries of the region and strives to dismantle the inner ones. People are called upon to 'stand up for their region' rather than slander it in contacts with external actors. People in general and politicians in particular are asked to 'pull together', to 'smoke pipes of peace', and to stop letting disunity impair the region's prospects for success. Besides, political actors are recurrently urged to work in networks and partnerships in which the 'joint regional interest' should come to the fore. In nationalist ideology, individuals are frequently called on to show loyalty to the

15 Dijkink (1996:12).

nation.[16] Yet nationalist ideologies tend to 'rest on the assumption of the relative superiority of *one* country and the relative inferiority of all others'.[17] It is the pre-eminence of one's own nation, one's own people and one's own territory that is emphasized in nationalism. In the regionalism we face here, it is not the region's superiority over other regions that is emphasized. Instead, it is 'regions in general' that are represented in essentially positive terms. We do find a wide range of policymakers stressing the good parts of their own region, and particularly in MV there are attempts to put forward the allegedly 'new' and 'unique' position of this region. The argument, however, is not that 'our' region is the best of all regions, but that the region needs to improve in order not to lag behind in global competition. In this striving, boundaries between interest groups within the region are either neglected or referred to as being problematic.

(b) History Formation

Region building is not only about the political use of boundaries, territory or space, or about a 'justification of the 'naturalness' of the territorial border'.[18] Region building is just as much about the political use of *time*. The importance of a selective writing of history in processes of region or nation building has been emphasized many times before.[19] The forgetting of history, a 'shared amnesia' or 'collective forgetfulness' has recurrently been identified as 'essential for the emergence of what we now consider to be a nation'.[20] Knowing that space could be 'nationalised through rewriting the past'[21], we need to be observant of attempts to 'regionalize' space by the same means. Thus, regional representatives of today certainly take 'liberties with history, seeking legitimacy

---

16 Kellas (1998:7).

17 Rejai (1995:54).

18 Dijkink (1996:12).

19 Renan, Ernst (1990), 'What is a Nation?', in Bhabha, H.K. (ed.), *Nation and Narration*, London:Routledge, p.52–53; Johansson, Rune (1993), 'Nationer och Nationalism: Teoretiska och Empiriska Aspekter', in Tägil, S. (ed.), *Den problematiska etniciteten*, Lund: Lund University Press; Tollebeek (1998:330); Ragin (1994:83).

20 Hobsbawm, E.J (1996), 'Ethnicity and Nationalism in Europe Today', in Balakhrishnan, Gopal (ed.) *Mapping the Nation*, London, New York:Verso, p.255; Gellner, Ernest (1995), *Culture, Identity, and Politics*, Cambridge: Cambridge University press, p.6.

21 Crang (1998:39).

in an imaged past' in pretty much the same ways as nation builders have proved to do before.[22]

In both regions, we find more or less explicit attempts to accentuate certain aspects of regional history.[23] In Norrbotten, we learn that there is a need to increase knowledge of the 'rich but neglected history' of the region. In Mecklenburg–Western Pomerania, we find several attempts to promote the region as an old cultural landscape. In both regions, public means have been used in campaigns intended to change the ways in which internal and external actors perceive the region. These campaigns could be understood as attempts to rewrite parts of the regional history. Hence, it is not unusual that features that have been perceived as negative in the past are presented as contemporary advantages or as something more or less neutral.[24] In Norrbotten, this is evident in references to interaction between the Swedish majority and the *Meänkieli* and Saami minorities. This interaction is recurrently referred to as a peaceful interrelationship that has increased the region's openness to cultural diversity in quite general terms. This representation of history, however, needs to be related to the fact that other voices have claimed that the two minorities have been subjected to both violence and discrimination over history.[25]

On the one hand, it is vital to consider the content of the written history. Thus, in the historical sciences, it is quite generally accepted that historical narratives are to be understood as the result of the desires and needs of the present, and not as neutral reflections on the past.[26] Writing history is acknowledged to be a selective process imbued with ideological considerations; therefore, *usefulness* is an important aspect of official historical accounts.[27] On the other hand, it must be observed that regional historiography in itself – no matter what historical narratives are forwarded – also contains aspects of usefulness. Therefore, the writing of regional history amounts not only to a search for a 'usable' past.[28] It is also an act that contributes to the conception of the region as an entity with a past of any kind at all.

---

22 Keating (1998:88).

23 See Bond *et al.* (2003: 379,387).

24 Bond *et al.* (2003:387).

25 See Chapter 3.

26 Johansson, Rune (1997), 'Ständig förändring eller orörlig historia?', in Andersson, L.M. and Zander, U. (eds), *In med historien.* Lund: Historiska media.

27 See Hylland-Eriksen (1999:51).

28 Keating (1998:84); Keating (2004:371).

In both regions, there is a tendency to provide the region with a past going back beyond the period in which it has formally existed. We encounter refereces to Norrbottnic culture as the result of 'thousands of years of interaction with nature', although the county as such has existed for two centuries only. In MV, interestingly, proclamations about the 'cultural roots' of the *Land* appear to leave out both much of the 40 years of state socialism and the Nazi era that preceded its constitution. Instead, they refer to a more distant past and to an alleged cultural affinity with Scandinavian countries. This 'extended past' can make a contemporary administrative region appear as being older, more 'genuine' and more 'natural' than it actually is. To locate a historical process or a past occurrence in an administrative unit that did not exist (either formally or in anyone's mind) when it took place could be understood as an attempt to provide that region with a new past, or to provide the past with a new territory. At all events, the region is provided with a historically continuous unity and specificity and, thus, the parallel to the use of history in processes of nation building is patent.[29] The impact of regional history may not be as great as the official writing of *national* history. Still, the mere endeavour to write the history of a contemporary, administrative region reflects an ambition to accentuate the outer boundaries of that region and an ambition to downplay or even erase internal conflicts.

(c) Community Building

Region building is not only about the political use of space and time, but can also 'reflect and reinforce ideas of what constitutes a people'.[30] We have seen that the 'regional population' is defined in quite similar ways in both regions. In some respects, it is those who have their roots in the region who stand out as ideal members of an alleged, regional community. This is particularly the case in Norrbotten. The 'Norrbothnian' is repeatedly adressed as having a certain mentality. This mentality is understood to be the result of a common 'Norrbottnic' experience of centuries of interaction with nature, wildlife and indigenous minority groups. Also in MV, we came across references to an alleged regional mentality. We found references to '*Subventionsmentalitäten*' and to a deeply rooted incapacity for entrepreneurship. In both cases, it was suggested that the region is not only an administrative entity, but also a cultural community kept together by a shared regional identity and by a particular, regional, way of life.

---

29 Tollebeek (1998:330).

30 Crang (1998:40).

The belief in the existence of distinct regional mentalities or 'ways of life' can easily be compared to 19th-century notions of national characters. These notions have – for good reason – been the target of severe criticism. Essentially, the maintenance of collective, territorial identities and mentalities is open to two objections. *First*, we must not presume that people who share the same history or the same cultural traditions necessarily share awareness of their togetherness or of their presumed collective being in the world. Hence, biological sex does not determine people's gender identities, and the material conditions under which people live 'do not alone determine class identities'.[31] And similarly, geographic or climatic conditions or circumstances do not alone determine regional or territorial identities. Put simply, people can have things in common but still not see the commonality. And people can live under similar conditions but still relate to these conditions in different ways. A *second* objection links up to Benedict Anderson's idea that nations are *imagined* communities rather than communities based on a common experience.[32] On such lines, we could conclude that a region like Norrbotten or Mecklenburg–Western Pomerania is far too big and far too heterogeneous to provide all its members with the same experience. On this logic, there is no such thing as a shared or common 'regional' experience, but possibly an image of such an experience.

In both regions, regional representatives have voiced a desire to strengthen the feelings of regional belonging among the population. This, together with the bid for 'authentic' cultural expressions and for preserving the allegedly unique cultural heritage certainly give this kind of regionalism an essentialist, excluding and traditional touch. The approach to culture and identies adopted in the two regions contributes, in addition, to the construction of cultural hierarchies. It contributes to an overemphasis on the value of allegedly traditional, place-bound and regional cultures, and it conduces to turning immigrant culture into something which should be 'tolerated' or 'accepted' but which, in any event, is far less rewarding. This implies that regionalism is not only a territorial ideology, resting on an 'emotional attachment to a place',[33] but an ideology that, like nationalism, 'stress[es] the cultural similarities of [its] adherents'.[34]

The region and the regional community are not only thought to be in need of consolidation, but are also represented as being in need of considerable renewal.

31 Abdelal R, Herrera, Y.M., Johnston, A.I, and McDermott, R. (2005), 'Identity as a Variable', Working Paper, Boston: Harvard University, p.4.

32 Andersson, B. (1991).

33 Dijkink (1996:11).

34 Hylland Eriksson, Thomas (1991), 'Ethnicity versus Nationalism', *Journal of Peace Research*, 28(3), pp. 263–278, p.264.

Old customs, habits and traditions are, accordingly, idealized only to a certain extent. The population is in both regions described as a bit too indolent, as a bit too whining, and as insufficiently interested in private enterprise. The stress on the alleged deficits of the regional population constitutes a palpable difference from how 'the people' has been approached in nationalism. Thus, in nationalism, there is a propensity to stress the alleged superiority of one's own people.[35] In both cases, one's 'own people' are thought of as having many shortcomings. The 'ideal' regional citizens are not necessarily those who are lineal descendants of the region, but those who have the will and the capacity to benefit the economic development of the region.

Clearly, the 'civicness' of this kind of regionalism has more to do with hopes for economic growth than with strivings for a deepened regional democracy. This, too, can have excluding effects, not only along lines of ethnic and cultural belonging but along lines of class and social status as well. Large-scale unemployment draws a sharp line between those who experience the region as a viable, promising labour market and those who are excluded from that same market. The focus on endogenous growth may also have dissimilar consequences in different parts of the regions. Accordingly, the civic, economistic account of the regional citizen is also based on an exclusionist logic. The emphasis on entrepreneurs, businessmen and educated specialists has very little to offer those who do not qualify as, or cannot identify with, such types.

The gender aspects of this representation of the regional community have remained so far unspecified. It would probably be inaccurate to say that the ideal citizens, as they stand out in the regional policy discourse, by necessity are *men*. Hence, when highly skilled experts or entrepreneurs are referred to in the regional policy discourse, they are normally neuters: that is, they are not referred to as being men or women. Nonetheless, there are aspects in the regionalism exposed that ought to be analysed from a gender perspective. To begin with, the idea of a common, regional interest contradicts the basic feminist idea that women are systematically subordinated to men, and that political action should aim to bring this subordination to an end. In neither of the regions is the idea of a joint, regional interest problematized or discussed from a gender perspective. Gendered power structures or patterns of gender segregation in the labour market are not in focus in the regional debate either.

Yet the image of the 'entrepreneur' has been subjected to feminist research, and so has working in regional partnership and networks. In these studies – still few in number – it has been established that images of entrepreneurs and busi-

35 Rejai (1995:54).

ness executives to a very large extent coalesce with traditional images of men.[36] It has also been demonstrated that women's organizations, normally weak in resources, have problems in asserting their interests against other, more influential groups included in the regional partnerships.[37] If we consider the value basis of regionalism, it can only be concluded that the regionalism we face here is *gender blind* rather than *gender neutral.* Explicitly, the awareness of what implications a certain policy has for men and women respectively seems to be rather weak. Given this, the gender aspects of regionalism need to be further elaborated in future research.

When the region and its inhabitants are defined, the question of *power* must be taken into consideration. Thus, every definition of a collective 'we' is politically loaded and contains 'an element of power'.[38] The central question in this matter is *whose* definition of 'us', 'our region' or 'our identity' is being put forward. We need to ask who places a certain 'identity narrative ... on the agenda'[39] and why this narrative becomes influential. Thus, the desire to institutionalize a notion of commonality within the region does contribute to the exclusion of those who fail to live up to the professed commonality.[40] In both cases studied, a number of people do not fulfil the requirements for an (imagined) citizenship in the (imagined) regional community.[41] Thus, the ideal regional citizen is equipped both with a certain origin and with a commitment to contribute to the economic development of the region. Immigration – national and international – implies that a considerable share of the regional population fails to match the cultural romantic pictures drawn in many regional policy documents. The social stratification within these regions, too, excludes those sections of the population that have only a nominal chance of ever qualifying as 'entrepreneurial' or 'highly qualified' rolemodels.

---

36 Pettersson, Katarina and Saarinen, Ylva (2005), 'Entreprenörer, motorer och andra hjältar – Manlig norm i svensk klusterpolitik', *Nordisk Samhällsgeografisk Tidskrift,* 39, pp.37–59.

37 Hudson and Rönnblom (2003:101).

38 Paasi (2003a); Häkli, Jouni and Paasi, Anssi (2003), 'Geography, Space and Identity', in Öhman, J.and Simonsen, K. (eds.), *Voices from the North – New Trends in Nordic Human Geography*, Aldershot: Ashgate, p.143.

39 Paasi (2003a:481); see also Bourdieu (1991:221).

40 Verdery, Katherine (1996), 'Whither "Nation" and "Nationalism"', in Balakrishnan, G (ed.), *Mapping the Nation*, London and New York: Verso, p.231.

41 See Painter (2000).

(d) Formal Institution Building

The image campaigns referred to in this work can be understood as examples of so-called 'flag waving exercises'.[42] These campaigns and the symbols and ceremonies they host are – just like in nationalist movements – used to provide the region-building aspiration with a 'definite shape and force'.[43] The symbols, the campaigns and the events organized for promoting the region provide the people with an opportunity to come together in order to express their solidarity with the region.[44]

What we ought to remember, however, is that the building of nations or states never has been only about signs, symbols, discourses or narratives. In studies of nations and nationalism, the modern state has certainly been understood as 'one of the most important agents of identification and categorization'.[45] And the state has, indeed, assumed the prerogative to 'name, to identify, to categorize, to state what is what and who is who'.[46] But the main role of the state was never to name, categorize or identify. And the prominence of the state has never been based on its capacity to make discursive or symbolic categorizations. Instead, the categorizations made by the state became important since the state has had other capacities as well. The state has had the capacity to establish formal structures and institutions, to which people must relate in their daily lifes and which have lent the category of the 'national' a certain material substance. The efficiency of nationalist categorizations has, in consequence, been very dependent on the capacity of states. It is partially due to this capacity that territorial states have turned out to be the 'principal incubators of nationalism'.[47]

At a regional level, we do not find the same institutions or the same capacity as we do at the level of the state. We find categorizations, 'flag-waving exercises', symbols and image campaigns. But these exercises are seldom backed up with a state-like institutional set-up. There are indeed substantial differences between Norrbotten and Mecklenburg–Western Pomerania in terms of formal authority but, notably, neither of these regions has 'the panoply of competences that the nation-state in its heyday possessed to pull together the domains of culture and identity, economic development and politics'.[48]

42 Billig (1995:10).

43 Breuilly (1994:64).

44 See Breuilly (1994:64); See also Rejai (1995:6).

45 Brubaker, Rogers and Cooper, Frederick (2000), 'Beyond "Identity"', *Theory and Society*, 29(1), pp.1–47, p.27.

46 Brubaker and Cooper (2000:15).

47 Murphy (2002:195).

48 Keating *et al.* (2003:21).

Therefore, we cannot expect the categorizations made at the regional level to have the same efficiency as the categorizations made by the central state.

Still, the region-building efforts identified are not without formal, institutional elements. Thus, the partnerships and the networks referred to are important institutions in the regional organizational landscape.[49] This is particularly the case in Norrbotten. Given the limited influence of a Swedish county council, networks and partnerships constitute an important base for the political debate about regional development issues and external relations.

Many of the partnerships and networks are built upon a highly normative basis. To begin with, all of them are based upon the normative judgement that working in partnership is essentially good.[50] In several cases, these bodies have been established simply in order to encourage intra-regional cooperation and common regional stances. Some of them, particularly in the Norrbottnic case, also demand a positive attitude towards regional interaction in the European policy process. Particularly in a northern Swedish context, this is a stance that could be contested. Thus, popular opposition towards Sweden's membership of the EU was noticeable in northern Sweden, and a greater involvement in EU affairs is not necessarily compatible with that position. Moreover, many of the partnerships and networks presuppose a positive attitute towards business-involvement in policy formation and agenda setting. This, too, is contestable, since it can be understood to challenge or to undermine the authority of popularly elected authorities.

In fact, this way of organizing politics *circumscribes* the opportunities for a change of ideological course in regional developmental work. Hence, the organizational form chosen is not neutral, but is itself capable of hindering visions, agendas and policy recommendations that conflict with its ideological basis. The networks and partnerships are thus a form of institutionalized ideology that restricts the participants' room for manoeuvre and, most likely, also their preferences. Not only ideology itself 'tells people what is possible and what is right and wrong'.[51] The formal structure through which ideology operates also implicitly makes these recommendations.

There are many discursive similarities between Norrbotten and Mecklenburg–Western Pomerania. In MV, however, we find a *Landtag* and a government that can be held accountable in public elections. The *Landtag*, furthermore, provides the region with a more neutral platform for political discussions. Notably, we find in MV more 'flagging' for partnerships at the subregional level than at the

49 Süssner (2003).

50 See also Hudson and Rönnblom (2003).

51 Paasi (1996a:29).

level of the *Land*. This may be linked to the fact that the subregional level has a weaker organizational structure than the *Land* level. Therefore, it may be believed that working in consensual partnerships is more common or more evident in regions with limited decision-making power or with a weak or unclear political organization.

The Significance of 'A Regional Age'.

The focus of this work has been on the value basis of regionalism. The possible causes and effects of this regionalism have been discussed to a lesser extent. Nonetheless, my *fourth* concluding argument concerns the general context in which regionalism emerges nowadays. Thus, the many discursive similarities between the two regions have been patent throughout the study. Therefore, the reasons for these similarities have been of a certain interest. Why do political actors in two separate and in many respects dissimilar regions tell the same story in their policy documents? And how does it come about that socialists and conservatives seem to convey such largely similar messages? The ambition of this study has not been to 'trace' the statements analysed, and accordingly I cannot say 'from where' these statements originate. Yet we can try to understand the general context in which the statements and narratives referred to have been produced. With that, we can try to capture the possible driving forces behind the kind of regionalism we face here.

One suggestion could be that we have entered a 'Regional Age' or, at least, an age in which a wide range of structural preconditions contribute to the formation of new regions and to the consolidation of existing ones. Such a suggestion could be relevant, particularly when one considers that many theorists have linked the emergence of *nationalism* to a certain stage of technological, political and economic development.[52] One of the most influential voices in this matter is probably Ernest Gellner, who has stressed the significance of 'a National Age' for the rise of nationalism. Still, regions are very far from having all the competencies that the modern state had at the time of the rise of modern nationalism. The regional project may be prominent, but it is far from as hegemonic as the national one was once. Besides, the nation, the state and the nation state have not left the stage of political life and popular identifications. These facts, and the fact that the regional question is far from new, must be kept in the forefront of our minds.

52 Hobsbawn (1998:20); Kellas (1998:77); Gellner (1995:15ff.); Häkli and Paasi (2003:142).

All this notwithstanding, we must also acknowledge that in recent years the regional question has been raised at time when the nation state no longer constitutes the 'natural' or self-evident frame of reference.[53] The present is also a time in which there is an economically driven interest in making feelings of regional belonging, regional identities and regional cultural traits more conspicuous. My *fourth* argument, however, is that not only changing technological, political and economic conditions have contributed to the kind of regionalism we face here. In addition, a certain *way of thinking* of regions and regional development has come to exert a considerable influence on regional decision makers.

There is, to begin with, much *benchmarking* going on at the regional level whereby developmental strategies are compared across regions. In these processes, there is an inclination to choose other regions 'as a model to follow or to reject'.[54] More precisely, there is a tendency among regional representatives to highlight 'northern Italy', the 'four motors of Europe' or, in a Swedish context, Gnosjö as successful ideals worth learning from. Notably, though, the EU as well as state representatives are very much involved in and supportive of the benchmarking, the search for 'good examples', and the attempts to find 'best practices'.

The academic world has also contributed to setting the tone in the regional debate in recent years. In most dimensions of regionalism, there is a palpable merging of research, politics and implementation. A wide range of scholars, not only from within the regional sciences, have taken an active part in promoting regional action. They have done so by persistently stressing that the nation-state is on its way out, and that the regional level is the one at which a number of important social, political and economic processes are going to take place in the future.[55] The academic focus on endogenous growth, on regional competitiveness, and on soft location factors has also emerged in colloboration with politicians and civil servants at different geographical levels. Topics such as regional identities, cultures and images have also enjoyed a growing scholarly interest. In much research, these features are still approached as soft location factors, that is, as 'things' that may benefit the regional economy. Regional identities are in this respect thought to bring 'self-confidence',[56] 'social capital',[57] and civic spirit to

53 For a comprehensive discussion, see Keating (1998). See also Cole and Loughlin (2003).

54 Dijkink (1996:12).

55 For a few examples, see Florida (1995); Giddens (1999); Fukuyama (1995); Berggren and Brulin (2002); Törnqvist (1998).

56 Nilsson and Brulin (1997:3,54); Lindeborg (1995:64).

regions – all of which are understood as essential ingredients for regional economic success. Therefore, regional representatives can recommend and support regional identity and image campaigns, with the argument (among others) that 'research shows' that image and identity are important for the economy.[58]

The amalgamation of research, politics and implementation is certainly not new. Still, the ways in which research is connected to other parts of society differ over time and, as Öhman maintains, 'among disciplines and countries'.[59] In a Swedish context, the 'marriage' between the regional sciences on the one hand, and politicians and planners on the other, is said to have led to 'a strong orientation towards planning and applied research'.[60] It appears, however, as if the ideological course in planning and in applied research has changed over the years. It appears, too, as if the 'marriage' has come to include representatives of the business community to a greater extent than before. Representatives of regions, states, the European Union, and the regional sciences can even be said to form part of a *thought collective.*[61] A thought collective is, in Fleck's sense, a group of actors who exchange ideas and are involved in thought-related interaction with each other. Through an intense exchange of ideas and thoughts, a *thought-style* is developed, which makes a common product out of individual thoughts. In our case, the thought-style in view concerns a certain way of thinking about regions and regional development. This thought-style is characterized by a certain 'cognitive mapping' and rests on the 'hegemony of one geographical imagination over the other'.[62] This particular thought-style, I would say, guides or steers regional and regionalist policy quite generally in a certain direction. The result is what Lovering has called a 'new cognitive and normative framework, drawing attention to some issues and policy objectives and away from others'.[63]

---

57 Florida (1995).

58 See Chapter 7. For a discussion of economic justifications for cultural development strategies, see Mayo (2000:111ff.).

59 Öhman, J. (2003), 'Geography, Local Planning and the Production of Space: A Swedish Context', in Öhman, J. and Simonsen, K. (eds.), *Voices from the North – New Trends in Nordic Human Geography*, Aldershot: Ashgate, p.103.

60 Öhman and Simonsen (2003). For a discussion of this issue in a Brittish context, see McCrone (1993:509).

61 Fleck (1997).

62 MacLeod and Jones (2001:679), quoting Gregory, Derek (1994), *Geographical Imaginations*, Oxford: Blackwell.

63 Lovering (1999:380).

We may experience a time when regions are both built in a strict or formal sense and when existing regions are reinvented and filled with a new meaning. And, as argued above, policymakers may quite generally be involved in a widespread but yet time-specific way of thinking about regions. In that respect, there is a policy diffusion going on that owes more to general, 'horizontal', trends than to place-specific ones. But it is not enough to say that a number of horizontal factors conduce to the rise of a certain kind of regionalism. Instead, vertical or place-specific circumstances must also be taken in consideration when one aims to understand social change. Thus, regionalism is not a static set of ideas, but is situated in space and assumes different shapes in different settings. A *fifth* argument, therefore, is that some regions may be more likely than others to become 'regionalist'.

We know, for instance, that in both rural and urban regions attempts are made to give the area an unmistakable 'brand'. We know also that in rural and urban regions alike efforts are made to link culture to economy either by supporting cultural tourism or by supporting more generally the cultural economy. What still needs to be investigated is whether in peripheral, sparsely populated or rural regions there is a greater inclination to stress the alleged exoticism and authenticity of the region. There are, I would say, some indications pointing in this direction.

First, regional representatives in urban environments and in city regions have at least a theoretical opportunity to depict their region as a multicultural 'melting pot' or as a cosmopolitan place. In some western European cities, the multicultural has even become a fetish, used for marketing purposes and in attempts to 'brand' the place. These image strategies can indeed be called in question and accused of striving to conceal inequities between the majority population and ethnic minorities within the region. Attempts to provide a place with an image of prosperity can, furthermore, seek to conceal increased impoverishment and other underlying difficulties.[64] Yet the mere attempt to brand a place as 'cosmopolitan' or as a meeting place for people from all over the world indicates that cultural interchange and cultural renewal is seen as 'a good thing'. We do not necessarily find the same essentialism or the same fixation on 'authentic' culture in city regions as we do (as demonstrated above) in peripheral or rural areas.

64 Harvey (1989:14).

Second, in peripheral regions in which cultural tourism is thought to play an increasing role in the regional economy, it may be believed that there is more to gain from an 'exotic' or distinctive image. Thus, unless a region can brand itself as *more* international, *more* multi-ethnic, or *more* of a melting pot than some others, immigrant or 'non-regional' culture seems to be ascribed much less value. We saw, indeed, multiculture being referred to in Norrbotten as a typical trait of the region. The multiculture aimed at is still very place-specific, essentialized, and represented as 'rooted' in the region.

Third, principles for political legitimacy can differ over space. Economic growth is, as argued by Gellner, one of the main principles for political legitimacy.[65] Governing parties, it is argued, can exercise social control and counter opposition by providing a gradual and continuous amelioration of people's material standards.[66] When there are no opportunities to increase the living standards of the general public, however, other mechanisms for solidarity and loyalty need to come into effect. One of these other mechanisms, Gellner has argued, is nationalism. This line of argumentation is worthy of consideration also in a regional context. Thus, 'economic growth' may turn out to be a key political objective in many regions. But in regions with a weak economy, where growth fails to arrive or does not contribute to any particular amelioration of people's material standards, it may be that other principles of legitimacy come to play a larger role in political life. In these contexts, regionalist 'flag waving' and attempts to create territorial solidarity or some sense of place-bound identity'[67] may come to the fore. Narratives about a common regional past, present and future, and about regional identity and belonging can then be significant as attempts to create the needed legitimacy. Thus, it may not only be that 'the circus succeeds even if the bread is lacking', as Harvey has suggested.[68] It can also be that it is exactly when the bread is lacking that the circus gets really important.

This being so, we may not only talk about the significance of a 'Regional Age' or about the role of policy diffusion. Instead, we need, too, to talk about 'regional regions'. In these 'regional regions', we may find a greater willingness to portray the region as 'authentic', as culturally homogenous, and as kept together by a shared, historically derived, regional identity. In these regions, we may find more attempts to build the region on an exclusionist basis. Notably, these 'regional regions' do not need to be known for hosting 'ethno-regionalist

---

65 Gellner (1999:44).
66 Gellner (1997:37).
67 Harvey (1989:14).
68 Harvey (1989:14).

movements',[69] for being 'internal colonies'[70] or for hosting any allegedly pre-existing 'cultural distinctiveness'.[71] As this study shows, arguments about place-bound cultures and identities can play a prominent role in administrative regions, too. Yet complementary comparative studies are needed before we can draw conclusions concerning the eventual difference between narratives about authenticity and genuineness in various types of regions.

FUTURE RESEARCH ON REGIONALISM AND REGION BUILDING

This study, focusing on the *value basis* of regionalism, has accentuated many of the objectionable sides of contemporary regionalism. In this respect, the study counterbalances those studies that seem to start out from the normative position that regions, regionalism and the striving for regional identities and images are 'good things' in and of themselves. The study also disputes any notion that regional identities might be more perennial and 'authentic' than national identities.[72] Hence, it has been persistently argued that regions and regional identities are as 'constructed' as are nations and national identities. It has also been demonstrated that contemporary regionalism is far from free of ethnic exclusion and democratic deficits. With that, the study also counteracts any prospensity to represent recent forms of regionalism as intrinsically more democratic, more civic and more inclusive than 'old' regionalism. Taken as a whole, this study advocates a critical turn in research on regions and regionalism, regional development and region building. It also motivates a number of future research questions.

A *first* group of questions is motivated by the fact that many regional representatives are so concerned with promoting the region, with 'putting it on the map' and with creating a positive and distinct image for it. We need, initially, to ask whether the ambition to put the region on the map has spillover effects on policy areas other than those linked to culture and tourism.[73] Are there examples where spatial planning or decisions regarding public investments in

69 Schmitt-Egner (2002:188).

70 Hechter (1975).

71 See Rokkan, Stein and Derek W. Urwin (1983), *Economy, Territory, Identity: Politics of West European Peripheries*, London: SAGE.

72 See for instance Lindeborg, Lisbeth (1999), 'Regional Deep Structures in the German Cultural Space', in Tägil, S. (ed.), *Regions in Central Europe. The Legacy of History*, London: Hurst & Company, p. 53.

73 See Chapter 7.

infrastructure or housing have been influenced by the desire to profile the region? We also need to look more closely at cases in which representatives of various types of overlapping or even rival regions strive to 'brand' their region at the same time. What forms does the struggle for identification take in these cases? It must, furthermore, be recognized that most forms of 'region branding' involve not only political actors but professional advertising and public relations agencies, too. The decision to 'brand' or to promote the region can certainly be a political one, but the implementation and shaping of these campaigns involve also professional, commercial agents. This, too, raises a series of questions. How influential are these commercial agencies in the contemporary construction of regional stereotypes? To what extent do these agencies contribute to the widely accepted notion that 'putting the region on the map' needs to be a prioritized field of action? Furthermore, are any strategies adopted in regions to consult the citizens about how they want 'their' region to be promoted?

A *second* set of questions emerges from the observation that working in networks and partnerships is highly valued. We need to know more about the consequences of this way of organizing regional developmental work for those included in the networks. Do participants in these networks and partnerships identify common regional interests to a greater extent than the general public? And do the mental maps of these representatives take other shapes than is the case for the public at large? We need, also, to explore how political ideas 'travel' within and between these networks. From where do representatives in these networks and partnerships get their ideas and notions? What sources of information are shared by the participants in the networks, and what sources of information are specific for some individuals, parties, or organized interest groups? We need, furthermore, to investigate the issues around which conflicts emerge, how power is distributed among representatives within the networks, and what influence the networks have on policy making in the regions as a whole. The aforementioned hypothesis – that working in consensus-seeking partnerships might be more common in regions with limited decision-making power – also needs to be empirically investigated.

A *third* series of questions is linked to the fact that the regionalism identified proves to be permeated with both 'civic' and 'romantic' categorizations. Hence, it has been established that civic and romantic dimensions of regionalism alike have a palpable exclusionist potential. We should probably not overstate the impact of the categorizations made by regional representatives on people's identifications. Thus, the extent to which categorizations forwarded by political entrepreneurs really influence the self-understanding of people is very much an open question.[74] Yet excluding categorization may well have other consequences than those linked to the self-understanding of the popluation at large. The civic categorizations – highlighting entrepreneurs and highly skilled experts – may conduce to a policy less advantageous for those with no education or with poor prospects of establishing a business of their own. The romantic arguments – stressing the role of allegedly 'authentic' culture for the regional economy – may affect priorities made in cultural politics, in economic policy or in strategic planning. Whatever the case, we still need to consider the regionalist categorizations made as part of an ideology of whose effects we need to know much more.

In responding to these questions, it would be commendable to adopt a more critical approach towards concepts such as regional identities cultures and images. It would be advisable to problematize concepts such as 'regional competitiveness', 'civic spirit' and 'human resources'. Political attempts to focus on 'regional' interests and 'regional' solidarity should also be more critically received. Thus, there might very well be an *ambition* to set aside political affiliations, ethnicity, class, gender or other bases for solidarity. And it might be that there is a *desire* for a policy that is of good for the whole region. But from a scholarly perspective it needs to be observed that regionalism is far from neutral towards ethnicity, class or gender. And it is not 'neutral' in relation to ideological dividing lines either. The kind of regionalism we face here is, rather, constructed around a number of categorizations and stereotypes. It also revolves around a 'ranking' of different cultures, identities and images, of mental dispositions, skills and knowledge, and of political affiliations and ways to organize political work. The role of scholars in this context, I would say, should be to critically examine the excluding tendencies, the categorizations, the stereotypes and the rankings made, rather than just accepting, legitimizing or reinforcing them.

74 See Brubaker and Cooper (2000:27); Paasi (1996a); Anderson, M. (1997); Breuilly (1994:277).

Abbreviations

CDU	Christian Democratic Union. *Christlich Demokratische Union.*
Drs	Parliamentary Publiction. *Drucksache.*
FDP	Free Democratic Party. *Freie Demokratische Partei.*
Lst Norrbotten	Norrbotten County Administrative Board. *Länsstyrelsen Norrbotten.*
MfABL	Ministry of Labour, Building and Regional Development. *Ministerium für Arbeit, Bau und Landesentwicklung.*
MfBWK	Ministry of Education, Science and Culture. *Ministerium für Bildung, Wissenschaft und Kultur.*
MfWi	Ministry of Economics. *Wirtschaftsministerium.*
MV	Mecklenburg-Western Pomerania. *Mecklenburg-Vorpommern.*
NIR	Norrbotten International Council. *Norrbottens Internationella Råd.*
NLL	Norrbotten County Council. *Norrbottens Läns Landsting.*
NNN	*Norddeutsche Neuste Nachrichten* (regional daily).
NSD	*Norrländska Socialdemokraten* (regional daily).
NSEO	North Sweden European Office.

OZ	*Ostsee-Zeitung* (regional daily).
PDS	Party of Democratic Socialism. *Partei des Demokratischen Sozialismus.*
RALA	Regional Association of Local Authorities. *Kommunförbundet Norrbotten.*
RGA	Regional Growth Agreement. *Regionalt Tillväxtavtal.*
RGP	Regional Growth Programme. *Regionalt Tillväxtprogram.*
RPV	Regional Planning Association. *Regionaler Planungsverband.*
SAP	Swedish Social Democratic Party. *Socialdemokratiska Arbetarepartiet.*
SPD	Social Democratic Party of Germany. *Sozialdemokratische Partei Deutschlands.*
StK	State Chancellery. *Staatskanzlei.*

Documentary Sources

SWEDISH SOURCES

Abrahamsson, Olov (2002), 'En fråga om solidaritet", *NSD*, 2002-03-14.

Andersson, Inge (2002), 'Anförande", *EuropaForum IV i Sundsvall*, 2002-06-05.

Backgård, Kenneth (1998a), 'Mot Nya Djärva Mål", *Hjärtslag*, Nr 4, Boden.

Backgård, K. (1998b), 'Samarbete – Fördel Boden', *Hjärtslag*, Nr 4, Boden .

Björnemark, Maria (1998), Vägvisaren, hur har den uppfattats av Norrbottens aktörer? Luleå:CUFS.

Centerpartiet Norrbotten (1998), *Tillsammans lyfter vi Norrbotten*, Länsprogram 1998-2002, Luleå.

Centerpartiet Norrbotten (2002a), Lika villkor för livskraft i hela Norrbotten – Länsprogram, Luleå.

Centerpartiet Norrbotten (2002b), *Valmanifest*, Luleå.

Centerpartiet Norrbotten and Västerbotten (2001), *Den Norra vreden. Gemensamt uttalande av distriktsstämmorna*, 2001-03-24, Piteå.

Dalarnas Länsstyrelse (1996), *Med förenade krafter mot år 2007*, Falun och Borlänge.

Eriksson, P. *et al.* (2004), 'Vi är stolta over Fattenborg', *NSD* 2004-04-21.

Eriksson, P.O (2002), 'Norrbottens möjligheter måste alltid framhållas', *NSD*, 2002-09-04.

Eriksson, P.O (2003a), 'Företagandets, samarbetets och möjligheternas län', in Lst Norrbotten (ed.), *Bilder av Norrbotten*, Luleå.

Eriksson, P.O (2003b), 'Företagandets, samarbetets och möjligheternas län', *NSD* 2003-09-11.

Eriksson, P.O (2005), *Sprid inte Myter!*, Lst Norrbotten: Luleå.

Eriksson, S., Olsson, L., Lennartson, L., Agnemar, A. and Svensson, R. (2004), 'Turismen blir allt viktigare för de små kommunerna', *NSD* 2004-09-06.

Eriksson, Stig and Haarala, Ove (2003), 'Politik för hela länet – mer än bara sin egen bakgård', *NSD* 2003-07-28.

Europaforum Norra Sverige (2001), Positionspapper om Europeiska Unionens framtida sammanhållningspolitik, Lycksele.

Europaforum Norra Sverige (2002a), *Inbjudan, Europaforum Norra Sverige IV*, Härnösand.

Europaforum Norra Sverige (2002b), *Förslag till fortsatt inriktning*, Sundsvall.

Europaforum Norra Sverige (2003a), *Deklaration från Europaforum Norra Sverige V*, Östersund.
Europaforum Norra Sverige (2003b), 'Gemensam politisk kraftsamling', at *www.europaforum.nu*, 2003-09-26.
Finansdepartementet (n.d), *Lissabonstrategin – VM i hållbar tillväxt*, Stockholm.
Folkpartiet Liberalerna Norrbotten (1998), *Ny kurs för Norrbotten. Handlingsprogram 1998-2002*, Luleå.
Folkpartiet Liberalerna Norrbotten (2002a), *Ett län för individer! Valmanifest*, Norrbotten.
Folkpartiet Liberalerna Norrbotten (2002b), *Vi tycker att du ska rösta på oss!*, Luleå.
Forsberg, R. (2004), 'Framtiden finns i Kalix', *NSD*, 2004-09-23.
Forsberg, Robert (2003), Samverkan är enda vägen', *NSD*, 2003-04-30.
Framtidslandet (2004), 'Intervju med Kari Marklund', at *www.framtidslandet.nu/index.asp*, 2004-05-19.
Fredriksson, Sture and Niva, Sören (2002), 'Samarbete istället för konfrontation', *NSD*, 2002-09-04.
Gröning-Degerlund, Lotta (2001a), 'Landstingspolitikerna ett sammansvetsat gäng', *NSD*, 2001-09-27.
Gröning-Degerlund, L. (2001b), 'Kultur stämplas som hobby', *NSD*, 2001-09-28.
Gröning-Degerlund, L. (2001c), 'Suck och stön ska sätt Norrbotten på kartan', *NSD*, 2001-10-05.
Gröning-Degerlund, L. (2002), 'Det behövs mer än ett konserthus', *NSD*, 2002-08-17.
Gröning-Degerlund, L. (2003a), 'Ta befälet, Eriksson', *NSD*, 2003-03-14.
Gröning-Degerlund, L. (2003b), 'Svar direkt', *NSD*, 2003-07-07.
Gröning-Degerlund, L. (2003c), 'Några arbetstips, P-O Eriksson', *NSD*, 2003-08-05.
Gröning-Degerlund, L. (2003d), '"Vi ids inte!" är fel inställning', *NSD*, 2003-08-06.
Gröning-Degerlund, L. (2003e), 'Dags för enighet och strategi', *NSD*, 2003-08-18.
Grönlund, Anna (2002), 'Jag vill måla om bilden av Norrbotten', *NSD*, 2002-10-10.
Grönlund, A. (2003a), 'Ryck upp er, politiker', *NSD*, 2003-05-14.
Grönlund, A. (2003b), 'Norrbotten kan växa och utvecklas', *NSD*, 2003-06-02.
Haparanda Kommun (2002), 'Många Tornedalingar på Gärdet', at *www.haparanda.se/haparandaklubben/gardet2003.htm*, 2004-05-19.
Hedberg, Göran (1998), 'Politik är ett Lagspel', *Hjärtslag*, Nr 5, Boden.

Hedkvist Petersen, Eva (2002), 'EU-geografin kan inte harmoniseras', *NSD*, 2002-06-08.
Hedström, J.O and Westerberg, J.O (1992), 'Förord', Lundin, Kerstin (ed.), *Norrbottens synliga historia: Norrbottens kulturmiljöprogram*, Luleå: Lst Norrbotten.
Henriksson, Jan (1998), Regional förlagsverksamhet och stöd till bok- och medieproduktion, Luleå: NLL.
Holmqvist, Per (1998a), 'Norrbottens Sjukvårdsparti – en Regional Kraft', *Hjärtslag*, Specialnummer, Boden.
Holmqvist, P. (1998b), 'Flitiga företagare i Övertorneå', *Hjärtslag*, Specialnummer, Boden.
Hultin, Bo (2003), 'Se möjligheter, inte svårigheter', *NSD*, 2003-01-02.
Hultin, B. (2004), 'Samverkan och positivt tänkande ger tillväxt', *NSD*, 2004-09-10.
Justitiedepartementet (2004), Nationella minoriteter och minoritetsspråk - En sammanfattning av regeringens minoritetspolitik, Ju 04.15: Stockholm.
Kommunförbundet Norrbotten (2000), *Tiderna förändrar oss*, Luleå.
Kommunförbundet Sörmland (1999), Strategi för länsövergripande samverkan i Internationella frågor, Sörmland.
Kultur Skåne (2004), *Utvecklingsplan för scenkonsten i Skåne*, Malmö: Kulturnämnden.
Landstingsförbundet Sverige (2003), 'Medlemmarnas internationella arbete" på *www.lf.svekom.se/int/*, 2003-10-28.
Lavander, Per and Öberg, Maria (2002), 'Nya utmaningar i Europas utkant', *NSD*, 2002-03-05.
Lindgren Ulf and Riklund Tomas (2004), *Norrbotten presenterar, slutrapport*, Luleå.
Lindström, Olle (2001), 'Ett nytt vägval krävs för att göra länet attraktivare', *NSD*, 2001-11-27.
Löfstedt, Marianne (2002), Rapport från rapportörsgruppen vid EuropaForum IV in Sundsvall, 2002-06-05.
Lööw, Börje (2002), 'Skippa skitsnacket och tala om utveckling', *NSD*, 2002-07-11.
Lööw, B. (2004), 'Så lockar vi investerare till Luleå', *NSD*, 2004-09-06.
Lst Norrbotten (1999a), Norrbotten – Mångfald och Kontraster, Luleå.
Lst Norrbotten (1999b), 'Min lilla bok om Norrbotten', *Pressmeddelande*, Luleå.
Lst Norrbotten (2000), 'Välkommen till Länsstyrelsens informationsprojekt om Norrbotten', at *www.norrbotten.se/infoproj/intro.htm*.
Lst Norrbotten (2001a), 'Länsrallyt i Norrbotten' at *www.bd.lst.se/news/tvavtal/010606.htm* Luleå, 2001-06-06.

Lst Norrbotten (2001b), *Internationellt Handlingsprogram*, Luleå.
Lst Norrbotten (2003a), *Fakta om Norrbotten,* Luleå.
Lst Norrbotten (2003b), *Perspectives on Norrbotten*, Luleå.
Lst Norrbotten (2003c), Norrbottens tillväxtprogram 2004-2007, Luleå.
Lst Norrbotten (2004a), *Facts about Norrbotten 2004*, Luleå.
Lst Norrbotten (2004b), Kompetensfrågor i samverkan – ett magasin om kompetensförsörjning, Luleå.
Lst Norrbotten and Norrbottens Museum (1998), *Kulturmiljövårdens idé*, Luleå.
Lst Norrbotten *et al.* (2000), *Samlat Program Dokument, Mål 1 Norra Norrland*, 2000-05-24, Luleå and Umeå.
Lst Värmlands län and Region Värmland (2003), *Tillväxtprogram för Värmland 2004–2007*, Karlstad.
Lst Västerbotten (2002), Norra Sverige i ett EU-perspektiv – Vitbok. Norrländska synpunkter på EU:s framtida strukturpolitik, Umeå.
Lst Västernorrland (2001), *Vision 2005 med perspektiv mot år 2010,* Västernorrland.
Marklund, Kari (2000), Redovisning av regionutvecklingsdagen på Pite havsbad 00-03-01, Luleå: Lst Norrbotten.
Marklund, Kari (2001a), 'Landshövdingen hälsar välkommen', at *www.bd.lst.se*, 2001-11-23.
Marklund, Kari (2002), 'Samspelet avgör den regionala utvecklingen', NSD, 2002-10-31.
Marklund, Kari (2003), 'Nio miljarder står på spel våren 2003', *NSD*, 2003-03-13.
Messing, Ulrica (2002a), 'Inledningsanförande' *Tillväxt 2002 – Sverige i framtiden*, Piteå.
Messing, Ulrica (2002b), 'Vi vill att hela Sverige ska leva och utvecklas', *NSD, 2002-07-06.*
Ministry of Justice (2003), 'National Minorities and Minority Languages', Fact *sheet*, Stockholm: Regeringskansliet, June 2003.
Nilsson, Frank (2003), 'Vitalisera välfärden', *NSD*, 2003-10-10.
Nilsson, Christer (2002), 'Anförande', *EuropaForum IV i Sundsvall*, 2002-06-05.
Nilsson, Frank (2002), 'Bra välfärd börjar i småföretagen', *NSD*, 2002-05-28
Nilsson, Jens (2002a), 'Inledningsanförande', *EuropaForum IV i Sundsvall*, 2002-06-05.
Nilsson, Jens (2002b), 'Anförande', *EuropaForum IV i Sundsvall*, 2002-06-06.
NIR (2003a), 'Minnesanteckningar' at *nir.norrbotten.se/prot/prot_2003-03-20.htm*, 2003-05-16.

NIR (2003b), 'Norrbottens internationella samarbete ska skapa resultat', *Pressinformation,* Luleå.
NIR (2003c), 'Rådets Uppgift' *at nir.norrbotten.se/syfte.htm*, 2003-10-28.
NIR (2003d), 'Ett nätverk för politiker', at *nir.norrbotten.se*, 2003-10-27.
NLL (1994a), Kulturen för Norrbotten framåt - Kulturpolitiskt program, Luleå.
NLL (1994b), *Landstingsplan 1995*, Luleå.
NLL (1995), *Landstingsplan 1996*, Luleå.
NLL (1996a), *Landstingsplan 1997*, Luleå.
NLL (1996b), Mediefakta 1996. Historik, Statistik och Utvecklingstendenser, Luleå.
NLL (1996c), Mediepolitiskt program, Luleå.
NLL (1997a), Vi arbetar för välfärd och utveckling, Luleå.
NLL (1997b), 'Dialog för ett bättre företagsklimat', *Dnr Lka1250-97*, Luleå.
NLL (1997c), *Landstingsplan 1998*, Luleå.
NLL (1998a), 'Angående Handlingsprogram för det internationella arbetet i Norrbottens län', *Dnr LK A1231-98*, Luleå.
NLL (1998b), Kultur ger livskraft och ett starkare Norrbotten, Luleå.
NLL (1998c), *Landstingsplan 1999*, Luleå.
NLL (1999), *Landstingsplan 2000*, Luleå.
NLL (2000a), Kultur ger Norrbotten kraft, Luleå.
NLL (2000b), *Landstingsplan 2001*, Luleå.
NLL (2002a), 'Yttrande över EU-kommissionens vitbok om styrelseformer i EU', *Dnr 677-02*, Luleå.
NLL (2002b), 'Vägvisaren, en välkänd strategi', *Pressinformation*, Luleå.
NLL (2002c), 'Norrbottens Internationella Råd bildat', *Pressinbjudan*, Luleå.
NLL (2004a), 'Statliga pengar till kultur och infrastruktur', *Pressmeddelande* 2004-03-29, Luleå.
NLL (2004b), 'Om Norrbottens Regionala Tillväxtavtal', at *www.nll.se/hg4.aspx?id=13985*, 2004-05-19.
Norrbotten Presenterar (2003), Norrbottensdagen på Gärdet - Exponeringserbjudande, Maj, Luleå.
Norrbottengruppen (1998), Internationell Strategi för Norrbottens län, Luleå.
Norrbottengruppen (1999), Regionalt Tillväxtavtal för Norrbottens län 2000-2002, Luleå.
Norrbottengruppen (2003), Norrbottens tillväxtprogram 2004-2007, 1:a utkast, 2003-04-30, Luleå.
Norrbottens Sjukvårdsparti (1998), *Regionalpolitik*, Norrbotten.
NSD (2001a), 'Företagsamheten spirar i Malmfälten', 2001-09-10.
NSD (2001b), 'De vill få slut på revirtänkandet, 2001-09-26.
NSD (2001c), 'Segelström varnar för Törnmans machokultur', 2001-11-09.

NSD (2001d), 'Arbetskraft till 15 000 jobb sökes', 2001-11-23.
NSD (2001e), 'Fattiga S-kusiner från landet mötes på kongress', 2001-11-26.
NSD (2002a), 'Det är jäkligt mycket gnäll', 2002-05-17.
NSD (2002b), 'Företagen får *en* röst', 2002-12-21.
NSD (2003a), 'Fester för sju miljoner, men utflyttarna vänder inte hem', 2003-06-12.
NSD (2003b), 'Landshövdingen vill ge en positiv bild av länet", 2003-06-14.
NSD (2003c), 'Jantelagen styr Norrbotten", 2003-08-13.
NSD (2003d), 'Slutfestat på Gärdet - kritiserat PR-projekt läggs ned", 2003-12-30.
NSD (2003e), 'Nordligt samarbete för EU-pengar", 2003-06-05.
NSD (2005), 'Saco-chef pekar ut norrbottningarna som arbetsskygga', 2005-04-22.
NSEO (1999), *Verksamhetsplan för år 1999-2000*, Bryssel, Luleå, Umeå.
NSEO (2000), *Verksamhetsplan för år 2000-2001*. Bryssel, Luleå, Umeå.
NSEO (2001), *Verksamhetsplan för år 2002*, Bryssel, Luleå, Umeå.
NSEO (2002a), *Verksamhetsberättelse för år 2002*, Bryssel, Luleå, Umeå.
NSEO (2002b), *Verksamhetsplan för år 2003*, Bryssel, Luleå, Umeå.
NSEO (2003a), 'About us', at www.northsweden.org/english/aboutns/aboutns.htm, 2003-08-19.
Näringsdepartementet (2002), *Rapport om tillväxtavtalen – andra året*, Ds 2002:34, Stockholm.
Öberg, Maria (2001), 'Står kirunapartiet och Törnman för utveckling?', *NSD*, 2001-10-16.
Öberg, M. (2002), 'Glesbygdskommunerna behöver samverkan', *NSD*, 2002-06-26.
Ohlsson, Staffan (1998a), 'Regionfullmäktige – Nödvändigt för demokratin', *Hjärtslag*, Nr 4, Boden.
Ohlsson, Staffan (1998b), 'Jantelagen', *Hjärtslag*, Nr. 4, Boden.
Oskarsson, Eva (1998), 'När partipiskan viner', *Hjärtslag*, Nr 4, Boden.
Östsam (2000), *Verksamhetsplan år 2001*, Östergötland.
Övertorneå Kommun (2002), 'Övertorneå i Göteborg 28/2 - 2/3 2002' at *http://www.overtornea.se/nyheter/goteborg.shtml, 2004-05-19.*
Partnerskap för Norrbotten (1995), Vägvisaren – Så här ska vi göra för att utveckla Norrbotten, Luleå.
Petersen, Karl (2001a), 'Kämpa för Norrbotten', *NSD*, 2001-11-06.
Petersen, Karl (2001b), 'Sluta gnälla, Lars Törnman', *NSD*, 2001-12-18.
Regionutredningen Norrbotten (1999a), Regionalt Samarbete i Norrbotten – en återblick. Delrapport 2, Luleå.

Regionutredningen Norrbotten (1999b), *Regionaliseringens drivkrafter. Delrapport 4*, Luleå.
Regionutredningen Norrbotten (2000a), *Seminarierapport 2000-01-27*, Luleå.
Regionutredningen Norrbotten (2000b), *Slutrapport*, Luleå.
Regionutredningen Norrbotten (2002), Region Norrbotten – samverkansorgan för Norrbottens län, Luleå.
Regionutredningen Norrbotten (2004), Handlingskraft! Delaktighet och Kraftsamling för Norrbotten, Luleå.
Rönnberg, Sten (2003), 'Journalister bör hålla sig goda för att fiska efter popularitet', *NSD*, 2003-08-11.
Rosengren, Björn (1998), 'Förord', Norrbottengruppen (eds.), *Internationell Strategi för Norrbottens län*, Luleå.
Rosengren, Björn (2000), 'Framåt för Norrbotten' at *www.norrbotten.se/arkiv/ artiklar/framot.htm*.
Rosengren, B. and Zakrisson, Kristina (2001), 'Norrbotten – Framtidslandet', *PT*, 2001-04-06.
Rosengren, B. (2003), 'Se på möjligheterna', *NSD Affärsliv*, 2003-10-17.
RTK Stockholm (2004), Västra Götaland och Skånes arbete med regionala utvecklingsfrågor. Samverkan med kommuner och andra aktörer, PM Nr 3:2004, Stockholm.
SAP Norrbotten (1997), Tillväxt i Norrbotten – av egen kraft, Regionalt partiprogram, Arvidsjaur.
SAP Norrbotten (2001) *Regionalt partiprogram*, Norrbotten .
SAP Norrbotten (2002) 'Tillsammans – för en bättre framtid i Norrbotten', *Regionalt Valmanifest*, Norrbotten.
SAP Norrbotten (2004a), 'De här åtgärderna vill vi se', *Uttalande från DS* 2004-02-03, Norrbotten.
SAP Norrbotten (2004b), 'Tillväxt och växtkraft i hela Norrbotten', *Uttalande från DS -kongressen 2004-07-27/28*, Norrbotten.
Ström, Mårten (2002), 'Luleå mot en ljus framtid', *NSD*, 2002-05-31.
Sundberg, Robert (2001), 'Ökad regional egoism missgynnar Norrbotten', *NSD*, 2001-11-24.
Sundkvist, Hans-Åke (2003), 'Vi vill inte ha bidrag, bara vår beskärda del', *NSD*, 2003-05-30.
Sundström, A. (2004), 'Norrbotten har fått en verktygslåda', *NSD* 2004-04-13
Sundström, Anders (2003), 'Socialdemokraterna har ett tungt ansvar för länet', *NSD*, 2003-03-21.
Thörnlund, Lennart and Henriksson, Gerda (2002), 'Kultur är utvecklingskraft', *NSD*, 2002-03-05.

Törnman, Lars and Backgård, Kenneth (2002), 'Norrbotten en tillväxtzon', *NSD*, 2002-06-10.
Vänsterpartiet Norrbotten (1998), *Länsprogram 1998-2002*, Luleå.
Vänsterpartiet Norrbotten (2001), *Motion om översyn av den samlade kulturbudgeten*, Dnr 774-00, Luleå.
Västmanland International Co-operation (2001), Internationell strategi och handlingsplan för Västmanlands internationella nätverk (VIN), Västerås.
Västra Götalandsregionen (2004), Regionalt tillväxtprogram för Västra Götaland 2004-2007, Göteborg.
Vikström, Inger (2000), *Slutrapport, Informationsprojektet*, Luleå: Lst Norrbotten.
Wennberg, Jenny (2003), 'Avskaffa Jantelagen!', *NSD*, 2003-06-23.
Westman, Bengt (2002), 'Fler regionalpolitiker behövs', *NSD*, 2002-06-14.
Westman, Bengt and Sundström, Anders (2002), 'Tornedalen kan bli en ny tillväxtregion', *NSD*, 2002-09-14.
Wikman, Lars (2003), 'Norrbotten – en framtida tillväxtregion eller ett län under avveckling?', *NSD*, 2003-08-09.
Wikström, Åsa (2003), 'Vart tog de vackra orden vägen?', *NSD*, 2003-10-15.
Wikström, Åsa (2004), 'Det måste löna sig att tro på sin idé', *NSD* 2004-09-25.
Zakrisson, Kristina (2000), 'Varje region behöver en egen tillväxtpolitik', *NSD*, 2000-09-28.
Zakrisson, Kristina (2003), 'Norrbotten behöver inga allmosor', *NSD*, 2003-05-28.

German Sources

ASP Landesbeirat (2002), Beschluss vom Umsetzungstand von Aktionsprogrammen aus dem ASP, 2002-06-18, Schwerin.
Bartels, Gerhard (2001), in *Plenarprotokolle Landtag MV*, 3/63 Sitzung 2001-06-28, Schwerin.
Blum, Andreas (1999), 'Neuen Herausforderungen Rechnung tragen, in PDS Fraktion Landtag MV (ed.), *Was macht Ihr da im Schloss*, Schwerin.
Borchart, Barbara (1999), in PDS Fraktion Landtag MV (ed.), *Was macht Ihr da im Schloss*, Schwerin.
Borchmann, Uwe (2004), 'Andauernde Diskussion um Hartz IV schadet sozialer Sicherheit', at *www.liberale-mv.de*, 2004-12-17.
CDU Landtagsfraktion MV (1996), *Identitätsgewinn im Aufbau Ost*, Schwerin

CDU Landtagsfraktion MV (2002), Abwandern oder Anpacken? Mecklenburg-Vorpommern – ein Land mit Perspektive!, Schwerin.
CDU MV (2001), Die Zukunftsoffensive der CDU: Politik gestalten – Mecklenburg-Vorpommern voran bringen, Schwerin.
CDU MV (2002), Ein neuer Wille fürs Land – Wahlprogramm der CDU, Schwerin.
Daedelow, Dietrich (2000), in RPV Meckleburgische Seenplatte (eds.), *REGIOnal*, No1, Neubrandenburg.
Delheim, Judith (2002), 'Regionales in Programdebatte und Praktischer politik der PDS', *Disput* 2.
Dettling Warnfried (2001), 'Perspektivenwechsel', Paper presented at *Denkwerkstatt 2020*, Berlin.
Drs.2/319 (1995-04-18), Kleine Anfrage der Abgeordneten Frau Schnoor, Schwerin.
Drs.2/386 (1995-05-12), *Antwort der Landesregierung*, Schwerin.
Drs.2/525 (1995-06-16), Kleine Anfrage der Abgeordneten Frau Lautenschläger, Schwerin.
Drs.2/608 (1995-07-11), *Antwort der Landesregierung,* Schwerin.
Drs.3/1212 (2000-03-29), *Antrag der Fraktion der CDU*, Schwerin.
Drs.3/2975 (2002-06-26), *Redebeitrag Minister Kauffold*, Schwerin.
Drs.4/227 (2003-02-14), *Antwort der Landesregierung*, Schwerin.
Ebenet, Otto (2004), 'Tourismuskonzeption für Mecklenburg-Vorpommern', at *www.investorenportal-mv.de/archiv/211.html*, 2004-06-29.
FDP MV (2002), 18 gute Gründe FDP zu wählen, Damit das Land nicht untergeht, Schwerin.
Finansministerium MV (2000), 'Einbringung des Haushaltsplan-Entwurfs 2001', *Pressemitteilung Nr 60/00*, Schwerin.
Grüne MV (2002a), Rasender Stillstand: Kulturpolitik in MV, 2002-02-04.
Grüne MV (2002c), Wirtschaftspolitisches Grundsatzpapier, 2002-11-23.
Grüne MV (2002d), Wirtschaftspolitische Leitlinien, 2002-11-24.
Grüne MV(2002b), Selbstbewußte ostdeutsche Perspektive, 2002-02-21.
Holter, H. (2000a), 'Die Osterweiterung: Chancen und Risiken für Mecklenburg-Vorpommern, *PDS Pressedienst Nr 51/52/2000*, Schwerin.
Holter, H. (2000b), in RPV Meckleburgische Seenplatte (eds.) *REGIOnal*, No 2, Juni 2000, Neubrandenburg.
Holter, H. (2003a), Aktive Arbeitsmarktpolitik statt gefährlicher Flickschusterei, Rede im Landtag 2003-05-22, Schwerin.
Holter, H. (2004a), 'Die Städte stärken - Wandel erfordert neue Allianzen', 2004-03-18, Berlin.

Holter, H. (2004b), 'Kräftige Impulse durch Initiative Baukultur im Land', *Grußwort zum Tag der Architektur 2004, 2004-06-24*, Rostock .
http://kampage.einfachanfangen.de/hintergrund/hintergr_idee.html, 2004-11-08.
http://kampage.einfachanfangen.de/kampagne/kamp.html, 2004-11-08.
Innenministerium MV (2003), *Burgernahe statt Bürokratie*, Schwerin.
Institut für Niederdeutsche Sprache (1999), 'Plattdeutsch jetzt anerkannt als Regionalsprache', at *www.ins-bremen.de/* 2004-10-07.
Kaufmann, Christoph (2000), in RPV Meckleburgische Seenplatte (eds.) *REGIOnal*, No 3, Neubrandenburg.
Kautz, M. (2000), in RPV Meckleburgische Seenplatte (eds.) *REGIOnal*, No 1, Neubrandenburg.
Kautz, M. (2000), in RPV Meckleburgische Seenplatte (eds.) *REGIOnal*, No 3, Neubrandenburg.
Klink, Harry (2005), FDP gegen Hochschulsterben - Hochschulen rechnen sich! PM vom 28.04.2005.
Kreher, Hans (2003), *Liberaler Aufbruch MV*, Schwerin.
Kulturbeirat MV (2002), Erarbeitung einer Kulturentwicklungskonzeption für das Land Mecklenburg-Vorpommern, April, Schwerin.
Kulturbeirat MV (2003), *Grundsatzpapier zur Kulturentwicklung im Land Mecklenburg-Vorpommern*, Schwerin.
Kulturforum MV (2002-02-26), *Ohne Titel*, Schwerin.
Landtag MV (2001a), *Bartels, Gerhard in Plenarprotokolle*, 3/63, 28. Juni, Schwerin.
Landtag MV (2001b), *Kauffold, Peter in Plenarprotokolle*, 3/63, 28. Juni, Schwerin.
Landtag MV (2001c), *Schnoor, Steffi in Plenarprotokolle*, 3/63, 28. Juni, Schwerin.
Landtag MV (2002a), *Schmidt, Karin in Plenarprotokolle*, 3/84, 26. Juni, Schwerin.
Landtag MV (2002b), *Holter, Helmut in Plenarprotokolle*, 3/81, 29. Mai, Schwerin.
Landtag MV (2002c), *Schnoor, Steffi in Plenarprotokoll*, 3/84, 26. Juni, Schwerin.
Landtag MV (2003a), *Gramkow, Angelika in Plenarprotokolle*, 4/13, 10. April, Schwerin.
Landtag MV (2003b), *Holter, Helmut in Plenarprotokolle*, 4/10, 12. März, Schwerin.
Landtag MV (2003c), *Lohse, Frank Ronald in Plenarprotokolle*, 4/17, 26. Juni, Schwerin.

Landtag MV (2003d), *Rehberg, Eckhardt in Plenarprotokolle*, 4/13, 10. April, Schwerin.
Landtag MV (2003e), *Voland, Angelika in Plenarprotokolle*, 4/17, 26. Juni, Schwerin.
Landtag MV (2004a), *Riemann, Wolfgang in Plenarprotokolle*, 4/31, 18. Februar, Schwerin
Landtag MV (2004b), 'Landtag MV tritt dem Olymp-Club bei', *Pressemittelung*, 2004-03-25.
LandtagsNachrichten MV (2002), *Quoting Bretschneider, Sylvia*, 2/2002, Schwerin.
LandtagsNachrichten MV (2003a), *Aus dem Plenum*, 8/2003, Schwerin.
LandtagsNachrichten MV (2003b), *Quoting Ringstorff, Harald*, 1/2003, Schwerin.
LandtagsNachrichten MV (2003c), *Quoting Schildt*, Ute, 6/2003, Schwerin.
LandtagsNachrichten MV (2004), Jubiläum, Vor zehn Jahren trat die Landesverfassung in Kraft, 1/2004, Schwerin.
Löhlein, B. (2000), in RPV Meckleburgische Seenplatte (eds.) *REGIOnal*, No 3, Neubrandenburg.
Löhlein, B. und Löhlein, G. (2000), in RPV Meckleburgische Seenplatte (eds.) *REGIOnal*, No 1, Neubrandenburg.
MfABL MV (2000a), 'Nachhaltige Politik für die Zukunft Mecklenburg-Vorpommerns', *Begrüßung der Teilnehmer der Denkwerkstatt 2020*, Parchim.
MfABL MV (2000b), 'Land braucht gutes Klima für junge Unternehmer, *Pressemitteilung Nr 105/00*, Schwerin.
MfABL MV (2000c), 'Kräftiger Impuls für Existensgründer', *Pressemitteilung Nr 222/00*, Schwerin.
MfABL MV (2000d), 'Unternehmer braucht das Land – nicht Unterlasser', *Pressemitteilung Nr 225/00*, Schwerin.
MfABL MV (2000e), Spatial Planning in Mecklenburg-Vorpommern, Schwerin.
MfABL MV (2000f), 'Vorpommern - Landesteil der besonderen Chancen', *Pressemitteilung Nr 99/00*, Schwerin.
MfABL MV (2001a), 'Denkwerkstatt – Plädoyer für Mittelkonzentration', *Pressemitteilung Nr 229/01*, Schwerin.
MfABL MV (2001b), *Denkwerkstatt 2020 – Chancen für Mecklenburg-Vorpommern*, Schwerin,
MfABL MV (2001c), Arbeitsmarkt- und Struktwentwicklungsprogramm, 2001-12-04, Schwerin.
MfABL MV (2002a), 'Kampagne für Gründer', *Pressemitteilung Nr 214/02*, Schwerin.

MfABL MV (2002b), Raumentwicklung im MV - Initiativen zur Regionalentwicklung, Schwerin.
MfABL MV (2002c), 'Für Arbeit und Lebensperspektiven – Plädoyer für eine strategische Allianz', *Rede des Minister Holter am treffen der Denkwerkstatt*, Wolgast.
MfABL MV (2003a), 'Wandern und Wiederkommen - eine Chance fur das Land', *Rede des Arbeitsminister Holter*, 2003-03-26, Schwerin.
MfABL MV (2003b), 'Start von Aktionsprogramm für Firmen, die neue Wege gehen', *Rede des Arbeitsminister Holter auf Regionalkonferenz Mecklenburgische Seenplatte*, 2003-04-24, Schwerin.
MfABL MV (2003c), 'Prominente unterstützen Initiative für Unternehmergeist', *Pressemitteilung Nr 56/03*, Schwerin.
MfABL MV (2003d), 'Unternehmergeist im Land vielfältig gefördert', *Pressemitteilung Nr 69/03*, Schwerin.
MfABL MV (2003e), 'Unternehmen sein ist toll – Unternehmen bleiben toller', *Pressemitteilung Nr 145/03*, Schwerin.
MfABL MV (2004a), 'Holter für eine Zukunft Ost in den Farben des Ostens', *Pressemitteilung Nr 24/04*, Schwerin.
MfABL MV (2004b), 'Das Untermehmen bringt Alle weiter', *OstseeZeitung, 2004-05-16.*
MfABL MV (2004c), 'Regionalmanagement und Regionalmarketing', at *www.am.mv-regierung.de/raumordnung/aktuell_regionalmanagement.html,* Schwerin, 2004-06-25.
MfBWK MV (2000), 'Kulturen im Dialog – gemeinsam gegen Gewalt', *Rede des Ministers Kauffold*, 2000-12-06, Schwerin.
MfBWK MV (2001a), 'Der Kulturaustausch der Völker im Ostseeraum' *Rede des Minister Kauffold auf Ostseeforum 2001*, Warnemünde.
MfBWK MV (2001b), 'Grußwort des Bildungsministers Prof. Peter Kauffold' *Pressemitteilung Nr 166-01*, Schwerin.
MfBWK MV (2002), 'Kulturentwicklung und kulturelle leitlinien für Mecklenburg-Vorpommern', *Rede des Ministers Kauffold*, 2002-06-05, Schwerin.
MfBWK MV (2003), Integratives Leitbild – Bildung Wissenschaft und Kultur für die Entwicklung von Mecklenburg-Vorpommern, 2003-12-03, Schwerin.
MfWi MV (1997), Kulturwirtschaft in Mecklenburg-Vorpommern, Schwerin.
MfWi MV (2003), Förderprogramme - Einzigartige Förderkulisse für Investoren in Mecklenburg-Vorpommern, Schwerin.
MfWi MV (2004), 'Starker Auftritt und klare Botschaft - MV tut gut', *Pressemitteilung Nr 42/04*, Schwerin.

MV4You (2005a), 'Projektbeschreibung', at *www.mv4you.de/project_ description.php*, 2005-05-19.
MV4You (2005b), 'Kluge Köpfe für MV', at *www.klugekoepfe4mv.de/uploads/flyer.pdf*, 2005-07-28.
Müller, Wolfgang (2003), 'Stadstumbau Ost', *Horisonte*, 2/2003.
NNN (2004), *Endspurt für Olympia*, 2004-05-16.
Olymp-Club Rostock (2003a), Bürger sollen Olympia-Bewerbung mitgestalten, 2003-12-04.
Olymp-Club Rostock (2003b), *Land gibt Regierungsgarantien für Olympia*, 2003-12-11.
OZ (2004a), 'Rostocker Senf passte nicht in Kühne-Konzept', 2004-02-06.
OZ (2004b), 'Wirtschaft stärkt Olympia-Bewerburg', 2004-05-12.
OZ (2004c), 'Various advertisements', 2004-05-16.
OZ (2004d), 'Ringstorff: Wir gehen mit gutem gefühl in diesen Tag ', 2004-05-18.
OZ (2004e), 'Aus der Traum', 2004-05-19.
OZ (2004f), 'Olymp-Club macht weiter und setzt sich neues Vereinsziel', 2004-05-19.
OZ (2004g), 'Party-Sturm fegte Tränen weg', 2004-05-19.
OZ (2004h), 'Rostock hat gewonnen', 2004-05-19.
PDS Landtagsfraktion MV (1999), *Was macht Ihr da im Schloss*, Schwerin.
PDS Landtagsfraktion MV (2003), 'Gemeinsam die Ausbildung von Jugendlichen sichern', *Presseerklärung*, 2003-05-28, Schwerin.
PDS Landtagsfraktion MV (2005), *Presseerklärung No 707*, 2005-04-19, Schwerin.
PDS MV (1997a), *Parteiprogramm des PDS 1998*, Schwerin.
PDS MV (1997b), *Wahlprogramm der PDS MV zu den Landtagswahlen 1998*, Schwerin.
PDS MV (2002), *Wahlprogramm 2002*, Neubrandenburg.
Pfeiffer, Ulrich, Dübel, Achim (1994), Grundzüge einer wirtschaftlichen Entwicklungsstrategie für Mecklenburg-Vorpommern, Friedrich-Ebert Stiftung Schwerin.
Pöker, Arno (2004), 'Gratulation an Mitbewerberstädte', *Pressemitteilung der Hansestadt Rostock*, 2004-05-19.
Region Rostock (2004a), *Rostock – Region der Chancen*, Rostock.
Region Rostock (2004b), *Kulturtouren*, Rostock.
Region Rostock (2004c), 'Regional Marketing Structure', at *www.regionrostock.de/Marketing_Initiative.267.0.html*, 2004-06-30.
Region Vorpommern e.V (no year), Vorpommern, Regionalmanagement, Regionalmarketing, Stralsund.

Rehberg, Eckhardt (2002), *Rede zu Wahlauftakt, 23 März 2002*, Warnemünde
Rettler, Bernhard (2000), in RPV Meckleburgische Seenplatte (eds.) *REGIOnal*, No 3, Neubrandenburg.
Ringstorff, Harald (2002), 'Vorwort', in StK MV (ed.) *Mecklenburg-Vorpommern – ein Land zum Leben*, Schwerin.
RPV Mecklenburgische Seenplatte (1998), *Regionales Raumordnungsprogramm*, Neubrandenburg.
RPV Mecklenburgische Seenplatte (2000a), *Regionales Entwicklungskonzept*, Neubrandenburg.
RPV Mecklenburgische Seenplatte (2000b), 'Regional Conference', at *www.region-seenplatte.de/html_e/page3.htm*, 2004-06-30.
RPV Mittleres Mecklenburg/Rostock (2001), 'Regionalmarketing, at *www.rpv-mmr.de/*, 2004-07-01.
RPV Mittleres Mecklenburg/Rostock (2002), *Regionale Agenda*, Rostock.
RPV Mittleres Mecklenburg/Rostock (2003), 'Regional Management' at *www.rpv-mmr.de/Regionalmanagement.19.0.html#74*, 2005-10-27.
RPV Westmecklenburg (2004a), 'Grüsswort von Regionalmanager Dr Jörg Köhn', at *www.westmecklenburg-schwerin.de*, 2004-06-25.
RPV Westmecklenburg (2004b), 'Regionalmanagements der Region Westmecklenburg' at *www.westmecklenburg-schwerin.de*, 2004-06-25.
RPV Vorpommern (1998), *Regionales Raumordnungsprogramm* Vorpommern, Greifswald.
RPV Vorpommern (2001), *Leitbild für die Region Vorpommern*, Greifswald.
RPV Vorpommern (2002), 'Aufgaben des Regionalmanagement – Vorpommern', at *www.rpv-vorpommern.de*, 2004-06-25.
Seite, Berndt (2002), 'Wohin, Mecklenburg-Vorpommern?', *Die Welt*, 2002-10-02.
Sinz, Manfred (2000), in RPV Meckleburgische Seenplatte (eds.) *REGIOnal*, No 2, Neubrandenburg.
SPD MV (2001), Bildung schafft Zukunft - Leitantrag des SPD Landesverbandes M-V, Güstrow.
SPD MV (2002), *SPD – Die Kraft des Landes*, Schwerin.
StK MV (1997), 'Werbekampagne für Mecklenburg-Vorpommern', *Presseinformation Nr. 115/97*, Schwerin.
StK MV (1998), Regierungserklärung zum Beginn der 3 Legislaturperiode, Schwerin.
StK MV (2000a), Es geht voran in Mecklenburg-Vorpommern, Schwerin
StK MV (2000b), Rede anlässlich des 10-jährigen Bestehens des Bundeslandes Mecklenburg-Vorpommern, 2000-10-18, Schwerin;
StK MV (2000c), *Entwicklungsbericht Vorpommern*, 2000-11-23, Schwerin

StK MV (2000d), Verfassung des Landes Mecklenburg-Vorpommern, Schwerin.
StK MV (2001), 'Bekämpfung der Arbeitslosigkeit ist größte gemeinsame Aufgabe', *Pressemitteilung Nr 93/01*, Schwerin.
StK MV (2002a), 'Chancen, Eigeninitiative, Zusammenhalt', *Rede, Minister Präsident Ringstorff auf FES-Tagung 2002-02-25*, Rostock
StK MV (2002b), 'Zusammenarbeit des Landes mit Südschweden soll weiter ausgebaut werden' *Pressemitteilung Nr 69*, Schwerin;
StK MV (2002c), Mecklenburg-Vorpommern – ein Land zum Leben, Schwerin
StK MV (2002d), 'Schwerpunkte der Landesregierung: Standortoffensive und Wirtschaftsförderung', *Rede anlässlich der Eröffnung des 3. Fördertages der Unternehmensverbände*, Schwerin.
StK MV (2002e), Vereinbarung SPD/PDS über die Bildung einer Koalitionsregierung für die 4. Legislaturperiode des Landtages von Mecklenburg-Vorpommern, Schwerin.
StK MV (2002f), Regierungserklärung zum Beginn der 4. Legislaturperiode, Schwerin.
StK MV (2002g), 'Kultur baut Brücken im Ostseeraum', *Rede des Minister President Ringstorff*, Schwerin
StK MV (2003a), 'Auswärtige Angelegenheiten', at *www.mv-regierung.de/stk/eu-abt/konaus.htm*, 2003-05-12, Schwerin.
StK MV (2003b), 'Mit Zuversicht und Selbstbewusstsein in die Zukunft sehen', *Pressemitteilung Nr 222/03*, Schwerin.
StK MV (2003c), 'Gemeinsam an einem Strang ziehen und Land weiter voranbringen', *Pressemitteilung Nr 137/03*, Schwerin.
Vater, Ulrich (2003), 'An einem Strang ziehen', *NNN*, 2003-11-19
www.am.mv-regieiung.de/ 2005-01-21
www.denkwerkstatt-mv.de/pages/start.htm, 2004-06-13
www.mv-tut-gut.de, 2004-10-07
www.nova-institut.de/modellregionen/, 2004-06-25

Bibliography

Abdelal R, Herrera, Y.M., Johnston, A.I, McDermott, R. (2005), 'Identity as a Variable', *Working Paper*, Boston: Harvard University.

Agnew, J., Livingstone, D. and Roger, A. (1996), *Human Geography: an Essential Anthology,* Oxford: Blackwell.

Ahonen, Pertti (2001), Soft Governance, Agile Union?, An Analysis of the Extensions of Open Coordination in 2000, Maastricht: European Institute of Public Administration.

Aldecoa, F. and Keating, M. (ed.) (1999), *Paradiplomacy in Action. The foreign relations of Subnational Governments*, London: Frank Cass.

Allen, J., Massey, D., and Cochrane, A. (1998), *Rethinking the Region*, London and New York: Routledge.

Althusser, Louis (1984), *Essays on Ideology*. London: Verso.

Alvesson, M and Sköldberg, K. (1994), Tolkning och Reflektion: vetenskapsfilosofi och kvalitativ metod, Lund: Studentlitteratur.

Amin, Ash (2000), 'Industrial Districts', in Sheppard E. and Barnes T. (eds.) *A Companion to Economic Geography*, Oxford: Blackwells.

Amin, Ash (2001), 'Moving On: Institutionalism in Economic Geography', *Environment and Planning A*, 33(1), pp.1237-1241.

Amin, Ash and Thrift, Nigel (1995), 'Institutional Issues for the European Regions', *Economy and Society*, 24(1), pp.41-66.

Anderson, Benedict (1991), *Imagined Communities,* London: Verso.

Anderson, J. and O'Dowd, L. (1999), 'Borders, Border Regions and Territoriality: Contradictory Meanings, Changing Significance', *Regional Studies*, 33(7), pp.593-604.

Anderson, J., O'Dowd, L. and Wilson, T.M. (2002), 'Why Study Borders Now?', *Regional and Federal Studies*, 12(4), pp.1-12.

Andersson, Lars M. and Zander, Ulf (1997) *In med Historien. Fem historiker om korta och långa perspektiv i historien*, Malmö: Lagerbringbiblioteket.

Anderson, Malcolm (1997), *Frontiers: Territory and State Formation in the Modern World*, Cambridge: Cambridge University Press.

Aronsson, P., Aspers, P., Beckman, L., Erixon, F., Micheletti, M., Möller, G., Strömberg, K., Uddhammar, E. (2001), *Dygder som drivkraft och föredöme*, Stockholm: City University Press.

Aronsson, Peter (1995), *Regionernas roll i Sveriges historia*, Stockholm: Fritzes.

Atkinson, Paul and Coffey, Amanda (2004), 'Analysing Documentary Realities', in Silverman, David (ed.), *Qualitative Research. Theory, Method and Practice*, London: SAGE.

Bacchi, Carol (2000), 'Policy as Discourse: what does it mean? Where does it get us?', *Discourse: Studies in the Cultural Politics of Education*, 21(1), pp.45-57.

Bahra, Hanne (1997), *Mecklenburg –Vorpommern*, Hamburg: Rowolth Verlag.

Ball, Terence and Dagger, Richard (1999), *Political Ideologies and the Democratic Ideal*, New York: Longman.

Baradat, Leon P. (1991), *Political Ideologies. Their Origin and Impact*, New Jersey: Prentice Hall.

Bastow, S., Martin, J., and Pels, D. (2002), 'Introduction: Third Ways in Political Ideologies', *Journal of Political Ideologies*, 7(3), pp.269-280.

Bauer, Lars (2001), *Politische Kultur in Mecklenburg-Vorpommern*, Rostock: Institut für Politik- und Verwaltungswissenschaften, Universität Rostock.

Beckman, Ludvig (2005). *Idéanalys – en introduktion till det kritiska studiet av politiska budskap*, Stockholm: Statsvetenskapliga Institutionen, Stockholms Universitet.

Bell, Daniel (1973), The Coming of Post-Industrial Society: A Venture in Social Forecasting, New York: Basic Books.

Berggren, Christian and Brulin, Göran (2002), Klistriga kluster eller globala glidare. Den lokala dynamikens paradoxala betydelse i den globaliserade ekonomin, Stockholm: NUTEK.

Betz, Hans-Georg (n. d.), 'Perspectives on Right-Wing Extremism', Canadian Centre for German and European Studies York University, Toronto at *www.aicgs.org/at-issue/ai-belz.shtml* 2004-03-28.

Billig, Michael (1995), *Banal Nationalism*, London and Thousand Oaks, CA: SAGE.

Bond, R., McCrone, D., Brown, A. (2003), 'National Identity and Economic Development: Reiteration, Recapture, Reinterpretation and Repudiation', *Nations and Nationalism* 9(3), pp.371-391.

Bondi, Liz (1996), 'Locating Identity Politics', in Keith, M. and Pile, S. (eds.), *Place and the Politics of Identity*, London and New York: Routledge.

Bort, Eberhard (2000), 'European Borders in Transition: The Internal and External Frontiers of the European Union', in *Boundary Permeability in Perspective, Papers Submitted to the Canadian Centre for Foreign Policy Development Conference*, University of Durham.

Boschma, R.A. (2005), 'Social Capital and Regional Development: an Empirical Analysis of the Third Italy', in: Boschma, R.A and Kloosterman, R. (eds.) *Learning from clusters. A Critical Assessment*, Dordrecht: Springer Verlag.

Bourdieu, Pierre (1991), *Language and Symbolic Power,* Cambridge: Polity & Basil Blackwell.

Brenner, Neil (2001), 'The Limits to Scale? Methodological Reflections on Scalar Construction', *Progress in Human Geography*, 25(4), pp.591-614.

Breuilly, John (1994), *Nationalism and the State*, Manchester: Manchester University Press.

Brown, David (2000), Contemporary Nationalism: Civic, Ethno-cultural and Multicultural Politics, London and New York: Routledge.

Brubaker, Rogers (1997), Nationalism Reframed. Nationhood and the National Question in New Europe. Cambridge: Cambridge University Press.

Brubaker, Rogers and Cooper, Frederick (2000), 'Beyond "identity"', *Theory and Society*, 29(1), pp.1 – 47.

Bryan, J, Hill, S, Munday, M, Roberts, A. (2000), 'Assessing the Role of the Arts and Cultural Industries in a Local Economy', *Environment and Planning A*, 32, pp.1391–1408.

Bräuninger, Michael (2002), *Social Capital and Regional Mobility*, Hamburg: Institut für Theoretische Volkwirtshaftslehre, Universität der Bunderswehr.

Burch, Martin and Gomez, Ricardo (2002), 'The English Regions and the European Union', *Regional Studies*, 36(7), pp.767–778.

Börzel, A. Tanja (2002), States and Regions in the European Union – Institutional Adaption in Germany and Spain, Cambridge: Cambridge University Press.

Calhoun, Craig (1994), 'Nationalism and Civil Society: Democracy, Diversity and Self-determination' in Calhoun, C. (ed.) *Social theory and the Politics of Identity*, Oxford: Blackwell Publishers.

Castensson, Reinhold (1994), 'Region – ett mångfaldigt begrepp', in Blomberg, B. and Lindquist, S.O (eds.), *Den Regionala Särarten*, Lund: Studentlitteratur.

Ceccato, Vania and Persson, Lars Olof (1999), *Forskning för regional utveckling*, Stockholm: Nordregio.

Chouliaraki, Lilie and Fairclough, Norman (1999), *Discourse in Late Modernity*, Edinburgh: Edinburgh University Press.

Christopoulos, Dimitrios (2001), 'Regional Elites and Public Entrepreneurship', Paper presented at *ECPR Joint Sessions of Workshops*, Grenoble.

Clement, Keith and Hansen, Malin (2001), *Sustainable Regional Development in the Nordic Countries*, Stockholm: Nordregio.

Cole, Alistair and Loughlin, John (2003), 'Beyond the Unitary state? Public Opinion, Political Institutions and Public Policy in Brittany', *Regional Studies*, 37(3), pp.265–276.

Colletis-Wahl, K. and Pecqueur, B. (2001), 'Territories, Development and Specific Resources: What Analytical Framework?', *Regional Studies* 35(5), pp.449–459.
Connor, Walker (2004), 'Nationalism and Political Illegitimacy', in Conversi, D. (ed.), *Ethnonationalism in the Contemporary World: Walker Connor and the Study of Nationalism*, London: Routledge.
Cooke, Phil (1997), 'Institutional Reflexivity and the Rise of the Region State' in Benko, G. and Strohmayer U. (eds), *Space and Social Theory: Interpreting Modernity and Postmodernity*, London: Blackwell.
Cooke, Phil and Leydesdorff, Loet (2005), 'Regional Development in the Knowledge-Based Economy: The Construction of Advantages', *Journal of Technology Transfer,* 30(3), forthcoming.
Crang, Mike (1998), *Cultural Geography*, London: Routledge.
Das, R. J. (2004), 'Social Capital and Poverty of Wage Labourers: Problems with the Social Capital Theory', *Transactions of the Institute of British Geographers*, 29(1), pp.27–45.
Deas, Iain and Ward, Kevin G. (2000), 'From the 'New Localism' to the 'New Regionalism'? The Implications of Regional Development Agencies for City-Regional Relations', *Political Geography* 19, pp.273–292.
Denk, Thomas (2002), *Komparativ metod – förståelse genom jämförelse*, Lund: Studentlitteratur.
Dien Hong (2005), *Vereinsgeschichte*, Rostock.
Dijkink, Gertjan (1996), *National Identity and Geopolitical Visions*, London: Routledge.
Doyle, Mike (2003), 'Discourses of Employability and Empowerment: Foundation Degrees and 'Third Way' discursive repertoires', *Discourse: Studies in the Cultural Politics of Education*, 24(3), pp.275–288.
EC (2000-03-23/24), 'Lisbon Presidency Conclusions', Lisbon.
Eckerberg, Katarina and Wide, Jessika (2000), *Actors, Institutions and Attitudes to Rural Development: The Swedish National Report*, Umeå: Department of Political Science, Umeå University.
Eisenstadt, S.N (2002), 'The Construction of Collective Identities and the continual reconstruction of Primordiality', in Malesevic, S and Haugaard, M. (eds.) *Making Sense of Collectivity*, London, Sterling, Virginia: Pluto Press
Ek, Richard (2003), *Öresundsregion - bli till!*, Lund: Lunds Universitet.
Ekman, Joakim (2001), *National Identity in Divided and Unified Germany*, Örebro: Örebro University.
Elenius, Lars (2001), *Både finsk och svensk*, Umeå: Umeå Universitet.
Erlandsson, Ulf (2005), 'Norrbotten', *Nationalencyklopedins internettjänst.*

Fairclough, Norman (1992), *Discourse and Social Change*, Cambridge: Polity Press, p.62
Fairclough, Norman (2003), *Analysing Discourse. Textual Analysis for Social Research*, London and New York: Routledge.
Fleck, Ludwik (1997), Uppkomsten och utvecklingen av ett vetenskapligt faktum. Stockholm/Stehag: Symposion.
Florida, Richard (1995), 'Toward the Learning Region', *Futures*, 27(5), pp.527–536.
Friedrich-Ebert Stiftung/Akademie Schwerin (1996), *Mecklenburg-Vorpommern, Land am Rand – für immer?*, Schwerin.
Fukuyama, Francis (1995), 'Social Capital and the Global Economy', *Foreign Affairs*, 74(5), pp. 89-103
Fukuyama, Francis (2004), 'The Imperative of State-building', *Journal of Democracy*, 15(2), pp.17–31.
Funderburk, Charles and Thobaben, Robert G. (1989), *Political Ideologies. Left, Center, Right*, New York: Harper & Row Publishers.
Galtung, Johan (1980), 'A Structural Theory of Imperialism', in Vogeler, I. and de Souza, A.R. (eds), *Dialectics of Third World Development*. New Jersey: Allanheld, Osmun & Co.
Garrison, Raymond M. and Massam, Bryan H (2001), 'Changing Governments and Changing Environmental Policies in Ontario: a discourse analysis', *Local Environment*, 6(3), pp.327–349.
Garsten, Christina and Jacobsson, Kerstin, eds. (2004), *Learning to be Employable. New Agendas on Work in a Globalizing World*, Basingstoke: Palgrave Macillan.
Gellner, Ernest (1983), *Nations and Nationalism*, Oxford: Blackwell.
Gellner, Ernest (1987), *Culture, Identity, and Politics*, Cambridge: Cambridge University Press.
Gellner, Ernest (1995), *Culture, Identity, and Politics*, Cambridge: Cambridge University Press.
Gellner, Ernest (1997), *Stat, Nation, Nationalism*, Nora: Nya Doxa.
Gellner, Ernest (1999), *Nationalism*, Nora: Nya Doxa.
Gertler, Meric, S. (2003), 'Tacit knowledge and the economic geography of context, or The undefinable tacitness of being (there)', *Journal of Economic Geography*, 3, pp.75–99.
Giddens, Anthony (1999), Tredje Vägen – om förnyelsen av socialdemokratin, Stockholm: Atlas.
Gidlund, Janerik and Sörlin, Sverker (1993), *Det Europeiska Kalejdoskopet*, Stockholm: SNS Förlag.

Gidlund J., Sörlin S., Gidlund S. (2000), *Ensam hemma - Den norrländska elitens nya syn på regional utveckling*, Umeå: CERUM, Umeå Universitet.
Glesbygdsverket (2002), 'Kommuner med den lägsta sysselsättningen. 2001. Andel sysselsatta. Procent' at *www.glesbygdsverket.se*
Glesbygdsverket (2004), 'Ohälsodagar per invånare fördelade på skogslänen och samtliga län. 16-64 år. 1999-2003. Antal.', at *www.glesbygdsverket.se*
Green, Anne E. and Turok, Ivan (2000), 'Employability, Adaptability and Flexibility: Changing Labour Market Prospects', *Regional Studies*, 34(7), pp.599–600.
Gregory, Derek (1994), *Geographical Imaginations*, Oxford: Blackwell.
Gregory, D., Martin, R. and Smith, G. (1994), *Human Geography*, London: Macmillan Press.
Gregson, Nicky, Simonsen, Kirsten and Vaiou, Dina (2001), 'Whose Economy for Whose Culture?', *Antipode*, 33(4), pp.616–646.
Gren, Jörgen (1999), The New Regionalism in the EU: the lessons to be drawn from Catalonia, Rhône-Alpes and West Sweden, Stockholm: Fritzes.
Gren, Jörgen (2002a), 'New Regionalism and West Sweden: The Factors of Change in the Regionalism Paradigm', *Regional and Federal Studies* , 12(3), pp.79 – 101.
Gren, Jörgen (2002b), *Den Perfekta Regionen*, Lund: Studentlitteratur.
Gruber, Karoly (2001), 'Regionalism, Nationalism, Integration; Central and Western European Perspectives', *The Challenge Europe On-Line Journal* 17(1).
Guibernau, Montserrat (1999), Nations Without States. Political Communities in a Global Age, Cambridge: Polity Press.
Gurgsdies, Erik (1996), 'Vom Rand in die Mitte? Wirtschaftsentwicklung in Mecklenburg' in Friedrich-Ebert Stiftung/Akademie Schwerin (1996) *Mecklenburg-Vorpommern, Land am Rand – für immer?*, Schwerin.
Haahr, Jens Henrik (2004), 'Open Co-ordination as Advanced Liberal Government', *Journal of European Public Policy*, 11(2), pp.209–230.
Hague, R., Harrop, M., Breslin, S. (eds.) (1992), *Comparative Government and Politics – An Introduction*, London: MacmiIllan Press.
Halkier, Henrik (2004), 'Institutions, Discourse and Regional Policy', *Lecture at ECPR Summer School*, Aalborg.
Hall, Stuart (1993), 'Rasism som ideologisk diskurs', *Häften för kritiska studier*, 3. Stockholm.
Hall, Stuart (1997), Representation: Cultural Representations and Signifying Practices, London: SAGE.
Hallin, G. and Malmberg, A. (1996), *Attraktion, konkurrens och regional dynamik i Europa*, Stockholm: Fritzes.

Hamilton, P. (2002), 'The Greening of Nationalism: Nationalising Nature in Europe', *Environmental Politics* 11(2), pp.27–48.
Hansen, Bernhard and Saglie, Jo (2005), 'Who Should Govern Political Parties? Organizational Values in Norwegian and Danish Political Parties', *Scandinavian Political Studies*, 28(1), pp.1–24.
Hansen, Peo (2000), Europeans Only? Essays on Identity Politics and the European Union. Umeå: Umeå University.
Harle, Vilho (1989), 'Towards a Comparative Study of Peace Ideas: Goals, Approaches and Problems', *Journal of Peace Research*, 26(4), pp.341–351.
Harmaakorpi, Vesa (2004), Building a Competitive Regional Innovative Environment – The regional Development Platform as a Tool For Regional Innovation Policy, Helsinki: Helsinki University of Technology.
Harvey, David (1989), 'From Managerialism to Entrepreneurialism: The transformation in urban governance in late capitalism', *Geografiska Annaler B*, 71(1), pp.3–17.
Harvey, David (1993), 'Class Relations, Social Justice and the Politics of Difference' in Keith, M. and Pile, S. (eds.), *Place and the Politics of Identity*, New York: Routledge.
Harvey, David (1996), Justice, Nature and the Geography of Difference, Oxford: Blackwell.
Harvie, Christopher (1994), *The Rise of Regional Europe*, London: Routledge.
Hechter, Michael (1975), *Internal Colonialism*, Berkeley, CA: University of California Press.
Hefeker, Carsten and Wunner, Norbert (2003), 'Promises Made, Promises Broken: A Political Economic Perspective on German Unification', *German Politics* 12(1), pp.109 – 134.
Held, David (1996), *Models of Democracy*, Cambridge: Polity Press.
Hellström, Anders (2003), '(Re)making European Unity', in Petersson, B. and Clark, E. (eds.), *Identity Dynamics and the Construction of Boundaries*, Lund: Nordic Academic Press.
Hill, Michael R. (1993), *Archival Strategies and Techniques*, Newbury Park, Calif.: SAGE.
Hobsbawm, E.J (1990), *Nations and Nationalism since 1780*, Cambridge: Cambridge University Press.
Hobsbawm, E.J (1996), 'Ethnicity and Nationalism in Europe Today', in Balakhrishnan, Gopal (ed.) *Mapping the Nation*, London and New York:Verso.
Hobsbawm, E.J and Ranger, T. (eds.) (1983), *The Invention of Tradition*, Cambridge: Cambridge University Press.
Hobsbawn, E.J (1998), *Nationer och nationalism*, Stockholm: Ordfronts Förlag.

Hoggart, K., Lees, L., and Davies, A. (2002), *Researching Human Geography*, London: Arnold.

Hogwood, Patricia and Gomez, Ricardo (2004), 'Implementing European Policy in Post Devolution Britain: Lessons from other EU Member States', *Manchester Papers in Politics* (11), Manchester: Department of Government, University of Manchester.

Hubbard, Phil (1996), 'Urban Design and City Regeneration: Social Representations of Entrepreneurial Landscapes', *Urban Studies,* 33(8), pp.1441–61.

Hudson, Christine (2005), 'Regional Development Partnerships in Sweden: Putting the Government Back in Governance?', *Regional and Federal Studies*, 15(3), pp.311–327.

Hudson, Christine and Rönnblom, Malin (2003), 'Uteslutande partnerskap?' in Håkan Brynielsson (ed.) *På jakt efter en ny regional samhällsordning!*, Stockholm: Svenska Kommunförbundet.

Hylland-Eriksen, Thomas, (1991), 'Ethnicity versus Nationalism', *Journal of Peace Research*, 28(3), pp.263–278.

Hylland-Eriksen, Thomas (1999), *Kulturterrorismen*, Nora: Nya Doxa.

Häkli, Jouni and Paasi, Anssi (2003), 'Geography, space and identity', in Öhman, J.and Simonsen, K. (eds.) *Voices from the North – New Trends in Nordic Human Geography*, Aldershot: Ashgate.

Institut für Niederdeutsche Sprache (1999), 'Plattdeutsch jetzt anerkannt als Regionalsprache', at *www.ins-bremen.de,* 2004-10-07.

Jacobsson, K. and Schmid, H. (2001), 'Real Integration or Just Formal Adaptation? - On the Implementation of the National Action Plans for Employment', Paper presented at *the Conference of the European Sociological Association*, Helsinki.

Jakubowski, Peter (2001), 'Regionen im Wettbewerb', *Informationen zur Raumordnung*, 8, Bonn: Bundesamt für Bauwesen und Raumordnung.

Jeffery, Charlie (2000), 'Sub-National Mobilization and European Integration: Does it Make any Difference?', *Journal of Common Market Studies* 38(1), pp.1–23.

Jensen, Ole B. and Richardsson, Tim (2004), *Making European Space. Mobility, power and territorial identity*, London and New York: Routledge.

Jerneck, Magnus (2003), 'Regional Lobbying och Flernivåstyrning', in Tallberg, P. (ed.), *Regionernas Europa*, Region Skåne och Västra Götalandsregionen.

Jerneck, Magnus and Gidlund, Janerik (2001), 'Region-building, Democracy and Changing Policy-Styles in Sweden', Paper presented at *Arena, Center for European Studies*, Oslo.

Johansson, Roine (2002), *Nyinstitutionalismen inom organisationsanalysen.* Lund: Studentlitteratur.

Johansson, Rune (1984), 'Varieties of Conflict Development', in Tägil, S. (ed.), *Regions in Upheaval*, Solna: Esselte Studium.

Johansson, Rune (1993), 'Nationer och Nationalism: Teoretiska och Empiriska Aspekter', in Tägil, S. (ed.), *Den problematiska etniciteten*, Lund: Lund University Press.

Johansson, Rune (1997), 'Ständig förändring eller orörlig historia?', in Andersson, L.M. and Zander, U. (eds), *In med historien*. Lund: Historiska media.

Johansson, Rune (1999), 'Ett odödligt sinnelag', in Olsson, Erik (ed.), *Etnicitetens Gränser och Mångfald*, Stockholm: Carlssons Bokförlag.

Jones, Martin (2001), 'The Rise of the Regional State in Economic Governance: 'Partnerships for Prosperity' or New Scales of State Power?', *Environment and Planning A*, 33, pp.1185–1211.

Jones, Martin and MacLeod, Gordon (1999), 'Towards a Regional Renaissance? Reconfiguring and Rescaling England's Economic Governance', *Transactions of the Institute of British Geographers*, 24(3), pp.295–313.

Jönsson, C., Jerneck, M., and Stenelo, L. (2001), *Politik i globaliseringens tid*, Lund: Studentlitteratur.

Keating, M., Loughlin, J. and Deschouwer, K. (2003), *Culture, Institutions and Economic Development*, Cheltenham, Northampton: Edward Elgar.

Keating, Michael (1997), 'The Political Economy of Regionalism' in Keating, M. and Loughlin, J (eds.), *The Political Economy of Regionalism*, London: Frank Cass.

Keating, Michael (1998), The New Regionalism in Western Europe – Territorial Restructuring and Political Change, Cheltenham UK, Northampton, MA USA: Edward Elgar.

Keating, Michael (2001), 'Rethinking the Region: Culture, Institutions and Economic Development in Catalonia and Galicia', *European Urban and regional Studies* 8(3), pp.217–234.

Keating, Michael (2004), 'European Integration and the Nationalities Question', *Politics and Society*, 32(3), pp.367–388.

Keating, Michael and Loughlin, John (2002), 'Territorial Policy Communities and Devolution in the United Kingdom', *EUI Working Papers, 2002/1,* Florence: European University Institute.

Keating, Michael and Loughlin, John (eds.) (1997), *The Political Economy of Regionalism*, London: Frank Cass.

Keating, Michael and McGarry, John (eds.) (2001), *Minority Nationalism and the Changing International Order*, Oxford: Oxford University Press.

Kellas, James G (1998), *The Politics of Nationalism and Ethnicity*, London: Macmillan Press.

Kim, Mi-Kyung and Robertson, John D. (2002), 'Analysing German Unification: State, Nation and the Quest for Political Community', *German Politics*, 11(3), pp.3–18.
Kockel, Ulrich (2002), Culture and Economy: Contemporary Perspectives, Aldershot: Ashgate.
Kohler-Koch, Beate (2002), 'European Networks and Ideas: Changing National Policies?', *EioP*, 6.
Krämer, Raimund (ed.) (1998), *Regionen in der Europäischen Union, Beiträge zur Debatte*, Berlin: Potsdamer Textbücher.
Kuhn, Heinrich-Christian (2000), 'Mecklenburg-Vorpommern – Gemeinsamkeiten trotz unterschiedlicher Geschichte', in Wehling, H.G (ed.), *Die Deutschen Länder – Geschichte, Politik, Wirtschaft*, Opladen: Leske+Budrich.
Kvist, Roger (1994), 'The Racist Legacy in Modern Swedish Saami Policy', Umeå: Department of Saami Studies, Umeå University.
Kymlicka, Will (1995), 'Misunderstanding Regionalism', *Dissent*, Winter, p.130–135
Kymlicka, Will (2000), 'Estonia's Integration Policies in a Comparative Perspective', in *Estonia's Integration Landscape: From Apathy to Harmony*. Tallinn: J.T Instituut, pp.29–57.
Landmann, Todd (2003), Issues and Methods in Comparative Politics, An Introduction, London and New York: Routledge.
Lapp, Peter Joachim (2000), 'Ein Land auf der Suche nach seiner Identität', *Aus Politik und Zeitgeschichte*, Nr. 37–38, pp.15–22.
Larsson, Reidar (1997a), *Jämförande politik*, Lund: Studentlitteratur.
Larsson, Reidar (1997b), *Politiska ideologier i vår tid*, Lund: Studentlitteratur.
Lash, Scott and Urry, John (2002), *Economies of Sign and Space*, London, Thousand Oaks, New Dehli: SAGE.
Le Galès, Patrick and Lequesne, Christian (eds.) (1998), *Regions in Europe*, New York: Routledge.
Lecours, André (2001), 'Political Institutions, Elites and Territorial Identity Formation in Belgium', *National Identities*, 3(1), pp.51–68.
Lewin, Leif (2002), *Bråka inte!*, Stockholm: SNS Förlag.
Liedman, Sven-Eric (1989), 'Om Ideologier', in Liedman, S.E. and Nilsson, I. (eds.), *Om Ideologi och Ideologianalys*, Inst. för Idé- och Lärdomshistoria, Göteborgs Universitet.
Lindeborg, Lisbeth (1995), *Regionalt Samarbete i Europa – med tyska erfarenheter*, Stockholm: Fritzes.
Lindeborg, Lisbeth (1999), 'Regional Deep Structures in the German Cultural Space', in Tägil, S. (ed.) *Regions in Central Europe. The Legacy of History*, London: Hurst & Company.

Lindeborg, Lisbeth (2003), 'Framgångsrika Regioner' in Tallberg, P. (ed.), *Regionernas Europa*, Region Skåne och Västra Götalandsregionen.
Lindström, Bjarne (1996), *Regional Policy and Territorial Supremacy*, Stockholm: NordREFO.
Lipset, Seymour (1981), 'The Revolt Against Modernity', in Torsvik P. (ed.), *Mobilisation, Center-Periphery-Structure and Nation-Building*, Bergen: Univ.-Forl., pp.451–500.
Living with Conflict (2003), 'Von Menschen, Ansichten und Gesetzen' at *www.living-with-conflict.org/v03/ cms.php?ID=602&language=DE,* 2003-11-21
Loughlin, John (2000), 'The Transformation of the State and New Territorial Politics', Paper presented at *Center for International Studies and Research*, Paris.
Lovering, John (1999), 'Theory Led by Policy: The Inadequacy of 'The New Regionalism', International Journal of Urban and Regional Research, 23(2), pp.379–395.
Lovering, John (2001), 'The Coming Regional Crisis (And How To Avoid It)', *Regional Studies*, 35(4), pp.349–354.
Lundén, Thomas and Zalamans, Dennis (2001), 'Local Co-Operation, Ethnic Diversity and State Territoriality – The Case of Haparanda and Tornio on the Sweden – Finland border', *GeoJournal*, 54(1), pp.33–42.
Lundquist, Lennart (1993), *Det vetenskapliga studiet av politik*, Lund: Studentlitteratur.
Lutz, Felix Philipp (2002), 'Historical Consciousness and the Changing of German Political Culture', *German Politics* 11(3), pp.19–34.
Löfgren, O. (1993a), 'Nationella Arenor', in Ehn, B., Frykman, J. and Löfgren, O. (eds.) *Försvenskningen av Sverige*, Stockholm: Natur och Kultur Förlag.
Löfgren Orvar (1993b), 'Det nationella i det globala' i Hemer, O. (ed.) *Kulturen i den globala byn*, Lund: AEGIS Förlag.
Löfgren, Orvar (1989), 'The Nationalization of Culture', *Ethnologia Europaea* XIX, pp.2–23.
Löfgren, Orvar (1994), 'Kurhotell, Pensionat och Sommarland' in Blomberg, B. and Lindquist, S.O (eds.), *Den Regionala Särarten*, Lund: Studentlitteratur.
MacKinnon, D., Cumbers, A. and Chapman, K. (2002), 'Learning, innovation and regional development: a critical appraisal of recent debates', *Progress in Human Geography* 23(3), pp.293–311.
MacLeod, Gordon (2000), 'The Learning Region in an Age of Austerity: Capitalising on Knowledge, Entrepreneurialism, and Reflexive Capitalism', *Geoforum* 31(2), pp.219–236.

MacLeod, Gordon (2001), 'New Regionalism Reconsidered: Globalization and the Remaking of Political Economic Space', *International Journal of Urban and Regional Research*, 25(4), pp.804–829.

MacLeod, Gordon and Jones, Martin (2001), 'Renewing the Geography of Regions', *Environment and Planning D,* 19(6), pp.669–695.

MacNeill, Donald and Tewdwr-Jones Mark (2003), 'Architecture, Banal Nationalism and Re-territorialisation', *International Journal of Urban and Regional Research*, 27(3), pp.783-43.

Maier, Jürgen (2003), 'Political Culture in East and West Germany', *Bamberger Beiträge zur Politikwissenschaft,* Nr. II–14,: Universität Bamberg.

Mankiw, Gregory N. (2004), *Principles of Economics*, Thomson South-Western.

Marks, G., Haesly, R., and Mbaye, H. (2002), 'What Do Subnational Offices Think They Are Doing in Brussels?', *Regional and Federal Studies*, 12(3), pp.1 – 23.

Marks, G., Nielsen, F., Ray, L. and Salk, J. (1996), 'Competencies, Cracks and Conflicts', in Marks, G., Scharpf, F.W., Schmitter, P. C., and Streeck, W. (eds.), *Governance in the European Union*, London: SAGE.

Markussen, Ann (1996), 'Sticky Places in Slippery Space: A Typology of Industrial Districts', *Economic Geography* 72(3), pp.293–313.

Marsh, David and Mackie, Tom (1995), 'The Comparative Method', in Marsh, D. and Stoker, G. (eds.), *Theory and Methods in Political Science*, London: Macmillan Press.

Marsh, David and Stoker, Gerry (eds.) (1995), *Theory and Methods in Political Science*, London: Macmillan Press.

Maskell, Peter and Törnqvist, Gunnar (1999), *Building a Cross-Border Learning Region*, Copenhagen: Copenhagen Business School Press.

Mayo, Marjorie (2000), *Cultures, Communities, Identities*, New York: Palgrave.

McCrone, David (1993), 'Regionalism and Constitutional Change in Scotland', *Regional Studies*, 27(6), pp.507–512.

McQuaid Ronald W. and Lindsay, Colin (2005), 'The Concept of Employability', *Urban Studies*, 42(2), pp.197–219.

Murphy, A.B. (2002), 'National Claims to Territory in the Modern State System: Geographical Considerations', *Geopolitics*, 7(2), pp.193–214.

Müller, D.K. and Pettersson, R. (2001), 'Access to Sami Tourism in Northern Sweden', Scandinavian Journal of Hospitality and Tourism, 1(1), pp.5–18.

Nairn, Tom (1997), *Faces of Nationalism*, London and New York: Verso.

Neumann, Iver B. (1994), 'A Region-Building Approach to Northern Europe', *Review of International Studies*, 20(1), pp.53–74.

Nilsson, Marianne and Brulin, Göran (1997), *Identiprenörskap - företagande med regionalt ursprung*, R 1997:22, Stockholm: NUTEK.

O'Loughlin, J., Witmer, F. and Ledwith, V. (2002), 'Location and Political Choice in Post-Unification Berlin', *Eurasian Geography and Economics* 43(5), pp.349–382.

Olsson, Jan and Åström, Joachim (2003), 'Why Regionalism in Sweden', *Regional and Federal Studies*, 13(3), pp.66–89.

Olzak, Susan (1998), 'Ethnic Protest in Core and Periphery States', *Ethnic and Racial Studies*, 21(2), pp.187–217.

Paasi, Anssi (1996a), *Territories, Boundaries and Consciousness*, New York: John Wiley & Sons Inc.

Paasi, Anssi (1996b), 'Regions as Social and Cultural Constructs: Reflections on Recent Geographical Debates', in Idvall, M, and Salomonsson, A. (eds.), *Att skapa en region – om identitet och territorium*, Stockholm: NordREFO.

Paasi, Anssi (2000), 'Re-constructing Regions and Regional Identity', *Lecture Manuscript, Nijmegen*, Netherlands.

Paasi, Anssi (2003a), 'Region and place: regional identity in question', *Progress in Human Geography* 27(4), pp.475–485.

Painter, Joe (2000), 'Multi-level Citizenship, Identity and Regions in Contemporary in Europe', in Anderson, J (ed.) *Transnational Democracy,* London: Routledge.

Papastergiadis, Nikos (2005), 'The Invasion Complex and the Globalisation of Fear', *Lecture at Dept. for Ethnic Studies*, Norrköping: Linköping University.

Peck, Jamie and Theodore, Nikolas (2000), 'Beyond "Employability"', *Cambridge Journal of Economics*, 24(6), pp.729–749

Peters, G.B (1998), *Comparative Politics, Theory and Methods*, Basingstoke: Macmillan Press.

Petersson, Bo and Clark, Eric (eds.) (2003), *Identity Dynamics and the Construction of Boundaries*, Lund: Nordic Academic Press.

Pettersson, Katarina and Saarinen, Ylva (2005), 'Entreprenörer, motorer och andra hjältar – Manlig norm i svensk klusterpolitik', *Nordisk Samhällsgeografisk Tidskrift,* 39, pp.37-59.

Pettersson, Örjan (1999), Population Changes in Rural Areas in Northern Sweden 1985-1995, Umeå: CERUM.

Pfeiffer, Ulrich and Dübel, Achim (1994), Grundzüge einer wirtschaftlichen Entwicklungsstrategie für Mecklenburg-Vorpommern, Schwerin: Friedrich-Ebert Stiftung.

Pin-Fat, Véronique (2005), 'The Metaphysics of the National Interest and the "Mysticism" of the Nation-State: Reading Hans J. Morgenthau', *Review of International Studies,* 31, pp.217–236.

Power, Dominique (2002), '"Cultural Industries" in Sweden: An Assessment of their Place in the Swedish Economy', *Economic Geography*, 78(2), pp.103–127.

Putnam, R. D. (1993), *Making Democracy Work: Civic Traditions in Modern Italy*, Princeton, NJ: Princeton University Press.

Ragin, Charles C. (1987), *The Comparative Method – Moving Beyond Qualitative and Quantitative Strategies*, Berkeley, Los Angeles, London: University of California Press.

Ragin, Charles C. (1994), *Constructing Social Research*, California: Pine Forge Press.

Ragin, Charles C. and Hein, Jeremy (1993), ' The Comparative Study of Ethnicity', in Stanfield, J.H and Dennis, R.M (eds.), *Race and Ethnicity in Research Methods*, Newbury Park, CA: SAGE.

Rehn, Alf (2004), 'Pessimistiskt Entreprenörsskap', *The Pink Machine Papers*, 1, Stockholm.

Reich, Konrad (ed.) (2001), *Mecklenburg-Vorpommern: ein Porträt*, Rostock: Konrad Reich Verlag.

Rejai, Mostafa (1995), *Political Ideologies, A Comparative Approach*, London and New York: M.E Sharpe.

Renan, Ernst (1990), 'What is a Nation?', in Bhabha, H.K., (ed.), *Nation and Narration*, London and New York: Routledge.

Rokkan, Stein and Derek W. Urwin. 1983. *Economy, Territory, Identity: Politics of West European Peripheries*. London: SAGE.

Rosing, Hans (1996), *Vetenskapens logiska grunder*, Helsingfors: Schildts förlag.

Rothstein, Bo (2000), 'Socialt Kapital i den socialdemokratiska staten. Den svenska modellen och det civila samhället', *Arkiv* nr 76.

Rumford, Chris (2002), *The European Union - A Political Sociology*, Oxford: Blackwell.

Rösler, Irmtraud (2004), 'Geschichte der Niederdeutsche Sprache', at *www.kulturportal-mv.de*, 2004-10-11.

Sack, Robert D. (1986), *Human Territoriality, Its Theory and History*, Cambridge: Cambridge University Press.

Sahlin, Ingrid (1999), 'Diskursanalys som sociologisk metod', in Sjöberg, K. (ed.), *Mer än kalla fakta. Kvalitativ forskning i praktiken*. Lund: Studentlitteratur.

Saïd, Edward W. (1993), *Orientalism*, Stockholm: Ordfronts Förlag.

Salomonsson, Anders (1994), 'Piteplat, Värmlandskorv och Äggakaka', in Blomberg, B. and Lindquist, S.O (eds.), *Den Regionala Särarten*, Lund: Studentlitteratur.

Sandemose, Aksel (1933), *En Flyktning Krysser Sitt Spor*, Oslo: Tiden Norsk Førlag.
Sartori, Giovanni (1970), 'Concept Misformation in Comparative Politics'. *American Political Science Review,* 64(4), pp.1033–53.
Sartori, Giovanni (1991), 'Comparing and Miscomparing', *Journal of Theoretical Politics*, 3(3): pp.243–257.
Sasse, Gwendolyn (2002), 'Conflict-Prevention in a Transition State: The Crimean Issue in Post-Soviet Ukraine', *Nationalism and Ethnic Politics,* 8(2), pp.1–26.
Scheinman, Lawrence (1977), 'The Interfaces of Regionalism in Western Europe: Brussels and the Peripheries', in Esman, M. J. (ed.), *Ethnic Conflict in the Western World*, Ithaca, New York: Cornell University Press.
Schirm, Stefan A. (2002), *Globalization and the New Regionalism*, Cambridge: Polity Press.
Schmitt-Egner, Peter (2002), 'The Concept of "Region": Theoretical and Methodological Notes on its Reconstruction', *European Integration*, 24(3), pp.179–200.
Schrijver, Frans J. (2003), 'Electoral Performance of Regionalist Parties and Perspectives on Regional Identity in French Regions', Paper presented at *ECPR Joint Sessions*, Edinburgh.
Schröder, Richard (1993), Deutschland schwierig Vaterland – Für eine neue politische Kultur, Freiburg: Herder.
Schröder, Richard (2000), 'Zehn Jahre deutsche Einigung', *Das Parlament,* 2000-09-29.
Scott, Allen J. (2000), *The Cultural Economy of Cities*, London: SAGE.
Seton-Watson, Hugh (1977), Nations and States. An Enquiry into the Origins of Nations and the Politics of Nationalism, London: Westview Press.
Silbergh, David M. (2001), *Doing Dissertations in Politics – A Students Guide*, London and New York: Routledge.
Skovdahl, Bernt (1992), *Tingsten, totalitarismen och ideologierna*, Stockholm: Symposion.
Smith, A.D (1990), 'The Suppression of Nationalism?', *International Journal of Comparative Sociology*, XXXI, 1–2.
Smith, A.D (1995), 'Gastronomy or Geology? The Role of Nationalism in the Reconstruction of Nations', *Nations and Nationalism*, 1(1), pp.3–23.
Statistische Bundesamt Deutschland, at *www.destatis.de/basis/e/bevoe /bev_bsp_t2.htm*, 2005-06-02.
Statistisches Landesamt at *www.statistik-mv.se*, 2005-06-02.
Statskontoret (1998), *Gamla län blir nya regioner?*, Stockholm.

Sweden's County Administrations (2005), 'The County Administration and the Community', at *www.lst.se/english/society.htm*, 2005–06-20.
Süssner, Josefina (2000), 'Regionen som identitetspolitiskt projekt', *Nordisk Samhällsgeografisk Tidskrift*, Nr 31.
Süssner, Josefina (2002), 'Culture, Identity and the Region', *Informationen zur Raumentwicklung*, Heft 4/5, Bonn: Bundesamt für Bauwesen und Raumordnung, pp.199-206.
Süssner, Josefina (2003), 'Nyregionalismens organisatoriska landskap', *Nordisk Samhällsgeografisk Tidskrift,* 35, pp.5–23.
Sörlin, Sverker (1988), *Framtidslandet*, Stockholm: Carlssons.
Taub, S., Piñar, P. and Galvan, D. (n.d.), 'Comparing Spatial Information in Speech/Gesture and Signed Language', Paper presented at *International Conference on Theoretical Issues in Sign Language Research*, Gallaudet University.
Taylor, Adam (2005), 'Bringing in the Norm: Non-ethnic, Non-Exemplar Regions', Paper presented at *Regional Identities, Cultures and Images – a Path to Regional Development?*, Norrköping.
Thompson, J.B. (1984), *Studies in the Theory of Ideology*, Cambridge: Polity Press.
Thompson, J.B. (1996), *Ideology and Modern Culture*, Cambridge: Polity Press.
Thompson, J.B., 'Introduction' in Bourdieu, Pierre (1991), *Language and Symbolic Power*, Cambridge: Polity & Basil Blackwell.
Tollebeek Jo (1998), 'Historical Representation and the Nation-State in Romantic Belgium (1830–1850)', *Journal of the History of Ideas* 59(2), pp.329–353.
Tomaney, John and Ward, Neil (2000), 'England and the "New Regionalism"', *Regional Studies*, 34(5), pp.471–478.
Tronvoll, Kjetil (1999), 'Borders of Violence – Boundaries of Identity: Demarcating the Eritrean Nation-state', *Ethnic and Racial Studies*, 22(6), pp.1037–1060.
Törnqvist, Gunnar (1993), *Sverige i Nätverkens Europa*, Malmö: Liber-Hermods.
Törnqvist, Gunnar (1998), *Renässans för Regioner*, Stockholm: SNS Förlag.
Ugalde Zubiri, Alexander (1999), 'The International Relations of Basque Nationalism and the first Basque Autonomous Government (1890-1939)'. in Aldecoa, F. and Keating, M. (eds.). *Paradiplomacy in Action: The Foreign Relations of Subnational Governments*, London: Frank Cass.
Ullrich, Melanie (2003), 'Definitions of Othering' at *www.cwrl.utexas.edu*, 2004-11-26.

Usher, A. (2004), A New Measuring Stick. Is Access to Higher Education in Canada More Equitable? Toronto: Educational Policy Institute.
Usher, A. and Cervenan, A. (2005), *Global Higher Education Rankings 2005*. Toronto: Educational Policy Institute.
van Dijk, Teun A (1985), 'Structures of News in the Press', in van Dijk (ed.), *Discourse and Communication*, Berlin: de Gruyter.
van Dijk, Teun A. (1993), *Elite Discourse and Racism*, California, London, New Delhi: SAGE.
van Dijk, Teun A. (1997a), 'The Study of Discourse', in van Dijk, Teun A. (ed.) *Discourse as Structure and Process. Discourse Studies: A Multidisciplinary Introduction*, Vol 1., London and Thousand Oaks: SAGE.
van Dijk, Teun A. (1997b). 'Discourse as Interaction in Society', in van Dijk, Teun A. (ed.), *Discourse as Social Interaction. Discourse Studies: A Multidisciplinary Introduction*, Vol. 2., London and Thousand Oaks: SAGE.
van Dijk, Teun A. (1998), *Ideology – a Multidisciplinary Approach*, London: SAGE.
van Dijk, Teun A. (2004), 'Politics, Ideology and Discourse', in Wodak, Ruth (ed.), *Encyclopedia of Language and Linguistics*, Barcelona: Universitat Pompeu Fabra.
Van Houtum, Henk and Strüver, Anke (2002), 'Borders, Strangers, Doors and Bridges', *Space and Polity*, 6(2), pp.141–146.
Weber, Max (1921), 'Politik als Beruf', *Gesammelte Politische Schriften*, München, pp.396–450.
Veggeland, Noralv (1994), Regionbygging over genser i det nye Europa – Aktörer, areaer or regimer, Stockholm: NordREFO.
Verdery, Katherine (1996), 'Whither 'Nation' and 'Nationalism'', in Balakrishnan, G (ed.) *Mapping the Nation*, London and New York: Verso.
Werz, Nikolaus (2000), 'Politische Reaktionen auf den Rechtsextremismus in Mecklenburg-Vorpommern', in Werz, N. and Hennecke, H.J (eds.) *Parteien und Politik in Mecklenburg-Voropmmern*, München: Olzog Verlag.
Werz, Nikolaus and Hennecke, Hans Jörg (eds.), (2000), *Parteien und Politik in Mecklenburg-Voropmmern*, München: Olzog Verlag.
Werz, Nikolaus and Schmidt, Jochen (eds.) (1998), *Mecklenburg-Vorpommern im Wandel – Bilanz und Ausblick*, München: Olzog Verlag.
White, Richard (1981), Inventing Australia: Images and Identity 1688-1980, Sydney: Allen & Unwin.
Wikhall, Maria (2003), 'Kultur och regional attraktivitet – migrationsbeslut bland högskoleutbildade i Sverige', in Egeland, H. and Johannisson, J. (eds.), *Kultur Plats Identitet*, Nora: Nya Doxa.

Williams, Colin H. (1997), 'Territory, Identity and Language', in Keating, M. and Loughlin, J. (eds.), *The Political Economy of Regionalism*, London: Frank Cass.

Williams, Raymond (1983), *Keywords*, New York: Oxford University Press.

Wollmann, Hellmut (2002), 'Local Government and Politics in East Germany', *German Politics*, 11(3), pp.153–178.

Öhman, (2003), 'Geography, local planning and the production of space: A Swedish context', in Öhman, J. and Simonsen, K. (eds.) *Voices from the North – New Trends in Nordic Human Geography*, Aldershot: Ashgate

Österberg, Eva (1994), 'Tradition och Konstruktion – Svenska regioner i ett historiskt perspektiv' in, Blomberg, B. and Lindquist, S.O. (eds.), *Den Regionala Särarten*, Lund: Studentlitteratur.

www.ingramcontent.com/pod-product-compliance
Ingram Content Group UK Ltd.
Pitfield, Milton Keynes, MK11 3LW, UK
UKHW021842100926
8556IPUK00019B/1008

9 783631 552018